Clinical Applications
of Capillary Electrophoresis

METHODS IN MOLECULAR MEDICINE™

John M. Walker, SERIES EDITOR

35. **Gene Therapy of Cancer:** *Methods and Protocols,* edited by *Wolfgang Walther and Ulrike Stein,* 2000

34. **Rotavirus Methods and Protocols,** edited by *James Gray and Ulrich Desselberger,* 2000

33. **Cytomegalovirus Protocols,** edited by *John Sinclair,* 2000

32. **Alzheimer's Disease:** *Methods and Protocols,* edited by *Nigel M. Hooper,*1999

31. **Hemostasis and Thrombosis Protocols:** *Methods in Molecular Medicine,* edited by *David J. Perry and K. John Pasi,* 1999

30. **Vascular Disease:** *Molecular Biology and Gene Therapy Protocols,* edited by *Andrew H. Baker,* 1999

29. **DNA Vaccines:** *Methods and Protocols,* edited by *Douglas B. Lowrie and Robert Whalen,* 1999

28. **Cytotoxic Drug Resistance Mechanisms,** edited by *Robert Brown and Uta Böger-Brown,* 1999

27. **Clinical Applications of Capillary Electrophoresis,** edited by *Stephen M. Palfrey,* 1999

26. **Quantitative PCR Protocols,** edited by *Bernd Kochanowski and Udo Reischl,* 1999

25. **Drug Targeting,** edited by *G. E. Francis and Cristina Delgado,* 1999

24. **Antiviral Methods and Protocols,** edited by *Derek Kinchington and Raymond F. Schinazi,* 1999

23. **Peptidomimetics Protocols,** edited by *Wieslaw M. Kazmierski,* 1999

22. **Neurodegeneration Methods and Protocols,** edited by *Jean Harry and Hugh A. Tilson,* 1999

21. **Adenovirus Methods and Protocols,** edited by *William S. M. Wold,* 1998

20. **Sexually Transmitted Diseases:** *Methods and Protocols,* edited by *Rosanna Peeling and P. Frederick Sparling,* 1999

19. **Hepatitis C Protocols,** edited by *Johnson Yiu-Nam Lau,* 1998

18. **Tissue Engineering,** edited by *Jeffrey R. Morgan and Martin L. Yarmush,* 1999

17. **HIV Protocols,** edited by *Nelson Michael and Jerome H. Kim,* 1999

16. **Clinical Applications of PCR,** edited by *Y. M. Dennis Lo,* 1998

15. **Molecular Bacteriology:** *Protocols and Clinical Applications,* edited by *Neil Woodford and Alan Johnson,*1998

14. **Tumor Marker Protocols,** edited by *Margaret Hanausek and Zbigniew Walaszek,* 1998

13. **Molecular Diagnosis of Infectious Diseases,** edited by *Udo Reischl,* 1998

12. **Diagnostic Virology Protocols,** edited by *John R. Stephenson and Alan Warnes,* 1998

11. **Therapeutic Application of Ribozymes,** edited by *Kevin J. Scanlon,* 1998

10. **Herpes Simplex Virus Protocols,** edited by *S. Moira Brown and Alasdair MacLean,* 1998

9. **Lectin Methods and Protocols,** edited by *Jonathan M. Rhodes and Jeremy D. Milton,* 1998

8. *Helicobacter pylori* **Protocols,** edited by *Christopher L. Clayton and Harry L. T. Mobley,* 1997

7. **Gene Therapy Protocols,** edited by *Paul D. Robbins,* 1997

METHODS IN MOLECULAR MEDICINE™

Clinical Applications of Capillary Electrophoresis

Edited by

Stephen M. Palfrey

Russells Hall Hospital, Dudley, West Midlands, UK

Humana Press ✳ Totowa, New Jersey

© 1999 Humana Press Inc.
999 Riverview Drive, Suite 208
Totowa, New Jersey 07512

This publication is printed on acid-free paper. ㊾
ANSI Z39.48-1984 (American Standards Institute) Permanence of Paper for Printed Library Materials.

Cover illustration: Fig. 2 from Chapter 1, "Clinical Applications of Capillary Electrophoresis," by Margaret A. Jenkins.

Cover design by Patricia F. Cleary.

For additional copies, pricing for bulk purchases, and/or information about other Humana titles, contact Humana at the above address or at any of the following numbers: Tel.: 973-256-1699; Fax: 973-256-8341; E-mail: humana@humanapr.com; Website: http://humanapress.com

Printed in the United States of America. 10 9 8 7 6 5 4 3 2 1

Library of Congress Cataloging in Publication Data

Main entry under title:

Methods in molecular medicine™.

Clinical applications of capillary electrophoresis / edited by Stephen M. Palfrey.
    p. cm. -- (Methods in molecular medicine™ ; 27)
    Includes bibliographical references and index.
    ISBN 0-89603-639-1 (alk. paper)
    1. Capillary electrophoresis. 2. Molecular diagnosis.
  I. Palfrey, Stephen M. II. Series.
    [DNLM: 1. Electrophoresis, capillary. 2. Proteins—analysis.
  3. DNA—analysis. QU 25 C641 1999]
  RB43.8.C36C55   1999
  616.07'56—21
  DNLM/DLC   98-32028
  for Library of Congress   CIP

Preface

The term "electrophoresis" was first used by Michaelis in 1909 to describe the migration of colloids in an electric field. The first practical electrophoresis method was described by Tiselius in 1937. He used a U-tube filled with buffer layered on top of sample; migration could be monitored using Schlieren optics. In zone electrophoresis, the U-tube was replaced by paper, a support material employed simply to prevent or minimize diffusion of ions, so that ions applied in a narrow strip to the paper will separate and remain as relatively discrete zones. Paper was superceded by a variety of other media, including cellulose acetate, hydrolyzed starch (starch gel), agarose, and polyacrylamide. The latter, in addition to being a support medium, has size-sieving properties. From the basic method of zone electrophoresis, other means of separation have been developed, including isoelectric focusing, isotachophoresis, density gradient electrophoresis, and various forms of immunoelectrophoresis.

In some ways Capillary Electrophoresis (CE) has gone full circle back to the original method of Tiselius. In its simplest form, separations occur in a buffer solution within a glass (fused silica) tube and detection occurs as sample moves past an optical window. CE has rapidly developed into a technique that rivals HPLC in its versatility. All the classical electrophoretic separations—zone, IEF, and isotachophoresis—have their counterparts in CE. Excitingly so, and authoritatively treated in *Clinical Applications of Capillary Electrophoresis*.

The addition of modifiers to electrophoresis buffers has opened up whole new separation possibilities. Adding detergents has created the technique known as Micellar Electrokinetic Capillary Chromatography (MECC); the method described in the chapter on steroids is a good example and offers an excellent description of the principles of the technique. Non-UV-absorbing species, such as anions and cations, can be detected by including chromophores in the buffer. In this instance detection occurs as a decrease in background absorbance as the ion passes the detector; detection of nitrate and oxalate use this method. Chiral separations can be achieved using additives like cyclodextrins, enabling the separation of optical isomers. One of the main driving forces of CE in fused silica capillaries is electroendosmosis; some additives like tetradecyltrimethylammonium bromide (used in the citrate/oxalate method) will reverse the endosmotic flow. Others such as methyl cellulose will reduce it greatly, an important consideration in the separation of hemoglobins by IEF.

Further modifications are possible by applying different coatings to the internal glass surface of the capillary. Some of these coatings, such as C18, will be familiar to HPLC users. The hemoglobin and CSF methods both use coated capillaries. Taking this process one step further, capillary electrochromatography (CEC) is a technique that combines CE and HPLC. Glass capillaries are filled with HPLC packing materials, the driving force being an electrical gradient rather than high pressure.

For the clinical laboratory, CE is potentially a very useful tool. It enables the automation of many methods that in the past were manual or semi-automated. In addition it offers great economy of reagent usage; assays can be run with as little as 100 µL or less of buffer at each end of the capillary. This has major advantages when expensive or hazardous organic solvents are used. It requires only nanoliter quantities of sample, allowing very small scale sample preparation where expensive reagents are used, such as in PCR. Consumables are restricted to sample vials and capillaries that, if looked after, should last months. Laboratory staff generally take to the technique well and find it reliable and robust. Compared with HPLC, startup time is minimal, and if something does go wrong, the capillary can be flushed out and ready for use in seconds. In addition it is capable of some extraordinary separations; peaks separated by only 6 s can be resolved to baseline and separations of up to a million plates/meter have been reported. Figure 4 in Chapter 19 shows this power clearly; seven peaks are resolved in less than half a minute, a separation efficiency of 400,000 plates/meter. On the negative side, CE can lack sensitivity when the optical path is equal to the capillary diameter and, at the moment, there is a limited range of detectors compared to those available for HPLC.

The range of CE assays available for the clinical laboratory has grown rapidly over the last two to three years. This book tries to give a representative, rather than comprehensive, look at those assays. Most laboratories will already carry out some form of electrophoresis, such as serum and urine proteins in clinical chemistry, or hemoglobins in hematology. Serum and urine protein analyses on CE use free solution electrophoresis with UV detection. The endpoint is an electropherogram that will be familiar to anyone who has used agarose electrophoresis with scan denitometry. The separation of hemoglobins uses isoelectric focusing with detection at 415 nm. This single assay allows both the detection and quantitation of abnormal hemoglobins together with HbA_2 and HbF, replacing several manual electrophoresis assays. Other methods describe the detection of CSF proteins, lipoproteins, myoglobin, cryoglobulins, HbA_1c, and cathepsin D.

One of the most rapidly expanding and exciting areas of CE is in DNA analysis. There are five chapters describing different aspects of this field; double-stranded DNA analysis, the prenatal diagnosis of Down's syndrome and Rh D/d genotyping, the identification of mutated p53 in cancers, the detection of microsatellite instability in cancers, and the detection of CMV. A significant advantage of CE is that it allows the injection of PCR products into the capillary without further sample cleanup.

Drug assays have long been performed in clinical laboratories, either as part of therapeutic drug monitoring (TDM) or as screening for drugs of abuse. TDM is frequently carried out by HPLC or immunoassay; CE offers a third way of performing these assays. The combination of the efficiency of CE separations coupled with diode array detection makes a powerful tool for the confirmation of the presence of drugs of abuse in urine.

Finally, *Clinical Applications of Capillary Electrophoresis* offers methods for some of the more esoteric analytes—nitrate and nitrite, oxalate and citrate, and serum polyamines—useful but not necessarily frequently requested assays.

In summary, *Clinical Applications of Capillary Electrophoresis* aims to give encouragement and guidance to laboratories new to CE and to demonstrate that the wide range of assays now available means that it is a technique tht has something to offer every clinical laboratory.

Stephen M. Palfrey

Contents

Preface v

Contributors xi

1 Clinical Applications of Capillary Electrophoresis
  Margaret A. Jenkins 1

2 Serum Protein Electrophoresis
  Margaret A. Jenkins 11

3 Urine Proteins
  Margaret A. Jenkins 21

4 Electrophoresis of Cerebrospinal Fluid
  Geoffrey Cowdrey, Maria Firth, and Gary Firth 29

5 Immunosubtraction as a Means of Typing Monoclonal
    and Other Proteins in Serum and Urine
  Stephen M. Palfrey 39

6 Analysis and Classification of Serum Cryoglobulins
  Zak K. Shihabi 47

7 Myoglobin Analysis
  Zak K. Shihabi 53

8 Enzyme Analysis: *Cathepsin D as an Example*
  Zak K. Shihabi 59

9 Quantification of Human Cytomegalovirus
    by Competitive PCR and Capillary Electrophoresis
  Zhongxin Yu, W. Douglas Scheer, and James M. Hempe 65

10 Laboratory Diagnosis of Structural Hemoglobinopathies
     and Thalassemias by Capillary Isoelectric Focusing
   James M. Hempe and Randall D. Craver 81

11 Serum Apolipoproteins
   Layle K. Watkins, Steven L. Cockrill, and Ronald D. Macfarlane 99

12 Gene Dosage in Capillary Electrophoresis: *Prenatal Diagnosis*
     of Down's Syndrome and Rh D/d Genotyping
   Pier Giorgio Righetti, Cecilia Gelfi, and Gian Franco Cossu 109

13 Rapid Analysis of Amplified Double-Stranded DNA
   by Capillary Electrophoresis with Laser-Induced
   Fluorescence Detection
   Ming-Sun Liu and Fu-Tai Albert Chen *121*

14 Identification of Mutated p53 in Cancers by Nongel-Sieving
   Capillary Electrophoretic SSCP Analysis
   Michiei Oto *127*

15 Detection of Microsatellite Instability in Cancers
   by Means of Nongel-Sieving Capillary Electrophoresis
   Michiei Oto *139*

16 Serum Lamotrigine Analysis
   Zak K. Shihabi *153*

17 Acetonitrile Stacking: *Serum Phenobarbital as an Example*
   Zak K. Shihabi *157*

18 Confirmation of the Presence of Drugs of Abuse in Urine
   Stephen M. Palfrey *165*

19 Steroid Analysis by Micellar Electrokinetic Capillary
   Chromatography
   ***Amin A. Mohammad, John R. Petersen,
   and Michael G. Bissell*** *177*

20 Determination of Polyamines by Capillary Electrophoresis
   Yin Fa Ma, Qingnan Yu, and Bingcheng Lin *189*

21 Urinary Oxalate and Citrate
   Ross P. Holmes and Martha Kennedy *199*

22 Plasma Nitrite and Nitrate Determination
   ***Toshiko Ueda, Tsuyoshi Maekawa,
   and Kazuyuki Nakamura*** *203*

Index *209*

Contributors

MICHAEL G. BISSELL • *Department of Pathology, Allegheny University of the Health Sciences; Department of Clinical Pathology, Allegheny General Hospital, Allegheny University Hospitals, Pittsburgh, PA*

FU-TAI ALBERT CHEN • *Beckman Instruments Inc., Fullerton, CA*

STEVEN L. COCKRILL • *Department of Chemistry, Texas A & M University, College Station, TX*

GIAN FRANCO COSSU • *Department of Immunohematology, Ospedale Civile di Sassari, Sassari, Italy*

GEOFFREY COWDREY • *Department of Biochemistry, Princess Royal Hospital, Haywards Heath, West Sussex, UK*

RANDALL D. CRAVER • *Department of Pathology, Louisiana State University School of Medicine and Children's Hospital, New Orleans, LA*

GARY FIRTH • *Princess Royal Hospital, Haywards Heath, West Sussex, UK*

MARIA FIRTH • *Princess Royal Hospital, Haywards Heath, West Sussex, UK*

CECILIA GELFI • *ITBA, CNR, Segrate, Milano, Italy*

JAMES M. HEMPE • *Department of Pediatrics, Louisiana State University School of Medicine, New Orleans LA*

ROSS P. HOLMES • *Department of Urology, Wake Forest University School of Medicine, Winston-Salem*

MARGARET A. JENKINS • *Division of Laboratory Medicine, Austin and Repatriation Medical Centre, Heidelberg, Victoria, Australia*

MARTHA KENNEDY • *Department of Urology, Wake Forest University School of Medicine, Winston-Salem, NC*

BINGCHENG LIN • *Department of Biochemistry, Truman State University, Kirksville, MO*

MING-SUN LIU • *Beckman Instruments Inc., Fullerton, CA*

YIN FA MA • *Department of Biochemistry, Truman State University, Kirksville, MO*

RONALD D. MACFARLANE • *Department of Chemistry, Texas A & M University, College Station, TX*

TSUYOSHI MAEKAWA • *Department of Emergency Medicine, Yamaguchi School of Medicine, Ube, Japan*

AMIN A. MOHAMMAD • *Departments of Pathology and Clinical Chemistry, University of Texas Medical Branch, Galveston, TX*

KAZUYUKI NAKAMURA • *Department of Biochemistry, Yamaguchi School of Medicine, Ube, Japan*

MICHIEI OTO • *Department of Biotechnology, Tokyo Technical College, Kunitachi-shi, Tokyo, Japan*

STEPHEN M. PALFREY • *Department of Clinical Biochemistry, Russells Hall Hospital, Dudley, West Midlands, UK*

JOHN R. PETERSON • *Departments of Pathology and Clinical Chemistry, University of Texas Medical Branch, Galveston, TX*

PIER GIORGIO RIGHETTI • *Department of Agricultural and Industrial Biotechnologies, University of Verona, Verona, Italy*

W. DOUGLAS SCHEER • *Department of Pediatrics, Louisiana State University Medical Center, New Orleans, LA*

ZAK K. SHIHABI • *Department of Pathology, The Bowman Gray Medical School, Wake Forest University, Winston-Salem, NC*

TOSHIKO UEDA • *Department of Anesthesiology-Resuscitology, Yamaguchi School of Medicine, Ube, Japan*

LAYLE K. WATKINS • *Department of Chemistry, Texas A & M University, College Station, TX*

QINGNAN YU • *The University of Texas, M.D. Anderson Cancer Center, Houston, TX*

ZHONGXIN YU • *Department of Pathology, Louisiana State University School of Medicine and Children's Hospital, New Orleans, LA*

1

Clinical Applications of Capillary Electrophoresis

Margaret A. Jenkins

1. Introduction

Capillary electrophoresis (CE) is a new and innovative technique that separates charged or uncharged molecules in a thin buffer-filled capillary by the application of a very high voltage. Separations by CE are extremely fast: Some are achieved in less than 5 min, with reproducibility studies often showing coefficient of variation (CVs) of <2%. The outstanding characteristic of CE is that it is an extremely sensitive technique. Early workers reported separations greater than 1 million theoretical plates per meter by CE, which is 10× the sensitivity of high-performance liquid chromatography (HPLC). The development of automated sample injection has meant that CE can be integrated into a clinical setting in which turnaround of accurate, cost-effective results are paramount.

Since 1937, when the original paper on electrophoresis by Tiselius was published *(1)*, many scientific papers have documented the progress of CE. Hjerten *(2)* originally suggested the usefulness of CE for zone electrophoresis and isoelectric focusing. Some excellent reviews on CE have already been published. Gordon *(3)* covered construction of instrumentation used in CE, as has Deyl *(4)*. Kuhr published a review of operational parameters and applications *(5)*. Mazzeo and Krull *(6)* reviewed coated capillaries for both capillary zone electrophoresis and capillary isoelectric focusing. In 1992, Shihabi *(7)* reviewed clinical applications of CE. Later, Jenkins et al. *(8)* and Lehmann et al. *(9)* also reviewed capillary electrophoresis applications in clinical chemistry.

2. Instrumentation

CE uses a very high voltage (1–30 kV) for the separation of analytes in the capillary, which may be either coated internally, or uncoated. Uncoated capil-

From: *Methods in Molecular Medicine, Vol 27: Clinical Applications of Capillary Electrophoresis*
Edited by: S. M. Palfrey © Humana Press Inc., Totowa, NJ

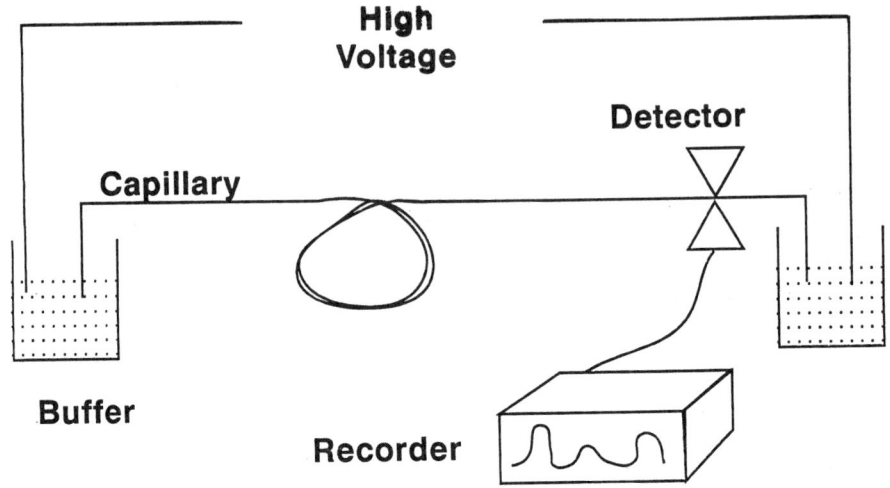

Fig. 1. Schematic diagram of CE apparatus.

laries are often referred to as fused silica capillaries. The diameter of the capillary varies between 20 and 100 μm in diameter, and is from 25 to 122 cm in length, depending on the configuration of the instrument. The ends of the capillary are placed in buffer vials, which also contain the electrodes. The narrow diameter of the capillary is important in heat dissipation from the high voltage applied, and also to decrease band diffusion. A schematic representation of a CE instrument is shown in **Fig. 1**.

Capillary columns have a polyimide outer covering, which makes the capillary mechanically stronger and protects the capillary from sudden angulation and breaking. The detector system with a CE may be variable wavelength, filter UV photometer, diode array, or a laser fluorescence detector *(10)*. At the detector window, the polyimide coating of the capillary is burnt off to allow the light source to penetrate the capillary, and for absorbance measurements of the analytes passing the window to be made.

Two methods are usually available for introduction of the sample to the capillary: electrokinetic and hydrodynamic injection. With electrokinetic injection, the inlet end of the capillary is removed from the buffer vial, and is inserted into the sample. A voltage is applied for a time ranging from 0.5–30 s, which causes the sample to migrate into the capillary. After the injection, the sample vial is replaced with the buffer vial and electrophoresis can proceed. The amount of sample introduced can be varied by altering both the time of injection and the injection voltage. The drawback of this type of injection is that sample components with highest electrophoretic mobility will be preferentially introduced over those with lower electrophoretic mobility *(11)*.

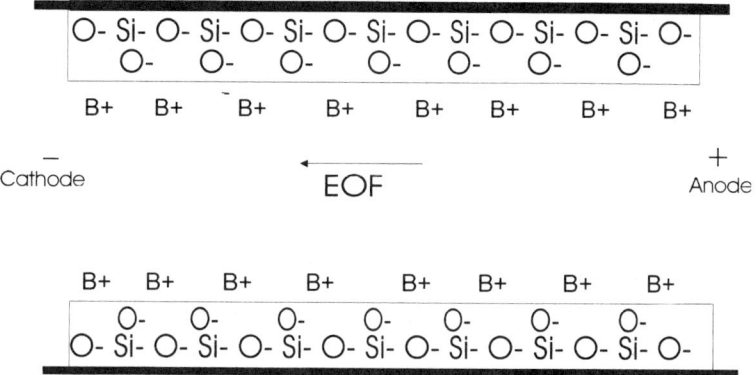

Fig. 2. Diagram showing cause of electroosmotic flow. Positively charged buffer ions, adjacent to the exposed negatively charged silanol ions of the fused silica wall, are attracted to the cathodes.

With hydrodynamic injection, the sample vial is raised above the capillary to a predetermined height, and the sample pours into the capillary for a defined period of time. Alternatively, the sample may remain at the same height as the outlet end of the capillary, and either positive pressure is applied to the sample vial or a vacuum is applied to the outlet electrode container. Hydrodynamic methods of sample introduction are not affected by the sample composition.

The type, pH, and ionic strength of the buffer are critical for the separations obtained *(12)*. Buffers may be made from a single component, such as phosphate, or may be quite complex, using two or more anions (borate–phosphate is a frequent combination). The pH of any buffer used in CE needs to be carefully optimized and maintained, to ensure reproducibility. The length of the capillary used, and the voltage applied, also influence the time of separation.

Electroosmotic flow is an important phenomenon in CE that can assist in the separation process *(13)*. The internal surface of fused-silica capillaries is negatively charged because of exposed silanol ions when the buffer is above pH 2.0. When an electric field is imposed, it causes hydrated ions in the diffuse double-layer adjacent to the silica wall to migrate toward the oppositely charged electrode, dragging solvent with the ions. This is termed electroosmotic flow, and can be used to advantage (*see* **Fig. 2**). The net flow of ions past the detector will reflect the balance between the electrophoretic and electroosmotic forces within the capillary. By adjusting the pH of the buffer in the capillary, electroosmotic flow can either enhance or oppose electrophoretic migration. Electroosmotic flow may also be decreased either by increasing the ionic strength of the buffer or by increasing the viscosity of the buffer by the addition of polymers, small amounts of organic solvents, or molecules such as glucose. Electroosmotic flow decreases with decreasing surface charge on the capillary, either by decreasing the pH of the buffer, or, alternatively, by decreasing the applied voltage *(14)*.

3. Modes of Separation

There are four major modes of separation by CE.

3.1. Capillary Zone Electrophoresis (CZE)

In free-solution capillary zone electrophoresis (CZE) a thin plug of sample is introduced into a buffer or gel-filled capillary. Under the influence of an external field, this yields discrete zones, which may be measured as they pass an in-line detector. Solutes are separated in this technique on the basis of differences in charge-to-mass ratio. In gel- or polymer-network-filled capillaries, solutes are separated, by the process of sieving, on the basis of their size.

Coatings for capillaries used in free-solution separations must be chemically stable and reproducible. For optimal separation, the surface modifications, which may be neutral or charged, should only partially inhibit electroosmotic flow. Examples of neutral coatings are polyacrylamide, methylcellulose, or polyethylene glycol. Charged coatings include quarternary ammonium functional groups bound to the capillary surface, or a small-mol wt polyethyleneimine coating that is suitable for basic proteins.

3.2. Isoelectric Focusing

Gel isolectricfocusing (IEF) can separate proteins that differ by as little as 0.001 of a pH unit *(15)*. As with gel IEF, capillary isoelectric focusing (CIEF) utilizes ampholytes that span the pH range of interest. These ampholytes facilitate high resolution separation of protein and peptide mixtures. CIEF usually uses a coated capillary; however, if the electroosmotic flow is sufficiently reduced by the use of methylcellulose or hydroxypropylmethylcellulose, then CIEF can be carried out in a fused-silica capillary. In CIEF, the capillary is filled with a mixture of protein sample and ampholytes. At the cathode, a basic solution (usually sodium hydroxide) is used, and an acidic solution (often phosphoric acid) is used at the anode. When an electric field is applied, the proteins migrate to the position at which the pH equals their respective pIs. When focusing has been completed, the current drops within the capillary to a minimal level.

Mobilization of peaks past the detector may be achieved by several methods. The first is electrophoretic mobilization, which involves adding salt to one of the electrolytes; for example, the addition of 80 mM NaCl to 20 mM NaOH *(16)*. Alternatively, mobilization of focused peaks may be achieved by the application of a vacuum to the capillary, as well as maintenance of the high voltage *(17,18)*. The third alternative is the recording of the pH gradient without mobilization. In practice, this is achieved by imaging the whole length of a short glass capillary *(19)*. Recently, cathodic mobilization has been achieved by replacing the catholyte with a proprietary zwitterionic solution (Bio-Rad, *20,21*), and by using gravity mobilization *(22)*.

3.3. Capillary Isotachophoresis

This mode of separation which employs stacking of dilute components, is not widely used, but in certain instances has useful applications. The stacking is achieved by using a small (e.g., 2 s) plug of water on either side of the injected sample. The result is that the sample becomes insulated from the buffer, resulting in sharper separations of dilute solutions.

An alternative isotachophoretic approach may involve using different leading and terminating electrolytes for focusing and preconcentration. After this step, the terminating electrolyte is replaced with the leading electrolyte for the remainder of the separation *(23)*.

3.4. Micellar Electrokinetic Capillary Chromatography

The essential characteristic of this type of separation, first described by Terabe in 1984 *(24)*, is the use of buffer containing surfactants at concentrations above their critical micelle concentration in fused silica capillaries. Thus micellar electrokinetic capillary chromatography (MECC) is a modification of CZE. Inside the capillary tube, there are two phases: a pseudostationary phase, which is an electrophoretically migrating micellar or slow moving phase, and an aqueous phase, with the electroosmotic force at a velocity higher than that of the micellar phase.

To be suitable for MECC, the micellar phase should be a surfactant that is highly soluble, and the solution must be UV-transparent and homogeneous. Examples of micellar systems include sodium dodecylsulphate (SDS), sodium deoxycholate, or SDS-tetra-alkylammonium micelles.

4. Clinical Separations

The number of scientific papers describing specific disorders diagnosed using CE has increased dramatically since 1990. These include DNA diagnosis of Down's syndrome *(25)*, adenylosuccinate lyase deficiency *(26)*, and P53 oncogene analysis *(27)*. Well-documented scientific papers are available on topics such as lipoprotein analysis by CE *(28)*, oxalate/citrate analysis *(29)*, plasma nitrate/nitrite *(30)*, organic acids in urine *(31,32)*, drugs of abuse in urine, anticonvulsants *(33)*, and urinary steroids *(34)*. CE techniques for serum proteins *(35–37)*, urine proteins *(38,39)*, hemoglobin variants *(40–41)*, cryoglobulins *(42)*, enzymes, cerebrospinal fluid (CSF) protein electrophoresis *(43)*, and HbA$_1$c *(44)* are also available.

Verification of a CE method, so that it can be introduced as a routine clinical method, involves testing at least 300 samples by both CE and conventional methods. The samples tested should be all the samples which come into the laboratory for that analyte, and must include at least 20% normal samples. If gross differences between the CE method and the conventional method are found during the testing of these 300 samples, then further samples should be assayed (up to 1000 samples) to show the proportion of these differently-behaving samples in everyday, routine testing. Statistical analysis of the two methods should be employed to show the correlation between the two methods, as well as the line of best fit.

The CE methods presented include the modes of free solution, IEF, micellar chromatography, and isotachophoresis.

One of the most informative aspects of this MIMM series is the Notes section, in which authors have indicated any problems or faults that can occur with their technique, and how these problems have been identified and overcome. This publication is aimed at scientists with no previous CE experience. The information contained within each chapter will allow validated methods to be successfully used by other laboratories keen to be involved with the rapid, sensitive, and extremely useful technique of capillary electrophoresis.

References

1. Tiselius, A. (1937) New apparatus for electrophoretic analysis of colloidal mixtures. *Trans. Faraday Soc.* **33,** 524–536.
2. Hjerten, S. (1990) Zone broadening in electrophoresis with special reference to high-performance electrophoresis in capillaries: an interplay between theory and practice. *Electrophoresis* **11,** 665–690.
3. Gordon, M. J., Huang X., Pentoney, S. L., Jr., and Zare, R. N. (1988) Capillary electrophoresis. *Science* **242,** 224–228.
4. Deyl, Z. and Struzinsky, R. (1991) Review capillary zone electrophoresis: its applicability and potential in biochemical analysis. *J. Chromatogr.* **569,** 63–122.
5. Kuhr, W. G. (1990) Capillary electrophoresis. *Anal. Chem.* **62,** 403R–413R.
6. Mazzeo, J. R. and Krull, I. S. (1991) Coated capillaries and additives for the separations of proteins by cpillary zone electrophoresis and capillary isoelectric focusing. *BioTechniques* **10,** 638–645.
7. Shihabi, Z. K. (1992) Clinical applications of capillary electrophoresis. *Ann. Clin. Lab. Sci.* **22,** 398–405.
8. Jenkins, M. A. and Guerin M. D. (1996) Capillary electrophoresis as a clinical tool. *J. Chromatogr. B.* **682,** 23–34.

9. Lehmann, R., Liebich, H. M., and Voelter, W. (1996) Application of capillary electrophoresis in clinical chemistry: developments from preliminary trials to routine analysis. *J. Cap. Electrophor.* **3,** 89–110.
10. Schwartz, H. E., Ulfelder, K. J., Chen F-T. A., and Pentoeny, S. L. Jr. (1994) Utility of laser-induced fluorescence detection in applications of capillary electrophoresis. *J. Cap. Electrophor.* **3,** 89–110.
11. Oda, R. P. and Landers, J. P. (1996) Introduction, in *Capillary Electrophoresis* (Landers, J. P., ed.), CRC, Boca Raton, FL, pp. 1–47.
12. McLaughlin, G. M., Nolan, J. A., Lindahl, J. L., Palmiere, R. H., Anderson, K. W., Morris, S. C., Morrison, J. A., and Bronzert, T. J. (1992) Pharmaceutical drug separations by HPCE: practical guidelines. *J. Liq. Chromatogr.* **15,** 961–1021.
13. Zhu, M., Rodriguez R., Hansen, D., and Wehr, T. (1990) Capillary electrophoresis of proteins under alkaline conditions. *J.Chromatogr.* **516,** 123–131.
14. El Rassi, Z. (1993) Capillary electrophoresis overview (theory and injection). Printed notes from workshop at *Conference on Capillary Electrophoresis,* Frederick.
15. Cornell, F. N. and McLachlan, R. (1985) Isoelectric focusing in the investigation of gammopathies, in *Clinical Biochemist Monograph,* Australian Association of Clinical Biochemists, Perth, pp. 31–37.
16. Zhu, M., Hansen, D. L., Burd, S., and Gannon, F. (1989) Factors affecting free zone electrophoresis and isoelectric focusing in capillary electrophoresis. *J. Chromatogr.* **480,** 311–319.
17. Chen, S-M. and Wiktorowicz, J. E. (1992) Isoelectric focusing by free solution capillary electrophoresis. *Anal. Biochem.* **206,** 84–90.
18. Chen, S.-M. and Wiktorowicz, J. E. (1993) High resolution full range (pI = 2.5 to 10.0) Isoelectric focusing of proteins and peptides in capillary electrophoresis, in *Techniques in Protein Chemistry 1V,* (Villafranca, J. J., ed.), Academic Press, San Diego, p. 333.
19. Wu, J. and Pawliszyn J. (1993) Fast analysis of proteins by isoelectric focusing performed in capillary array detected with concentrated gradient imaging system. *Electrophoresis* **14,** 469–474.
20. Zhu, M., Wehr, T., Levi, V., Rodriguez, R., Shiffer, K., and Cao, Z. A. (1993) Capillary electrophoresis of abnormal haemoglobins associated with alpha-thalassemias. *J. Chromatogr. A* **652,** 119–129.
21. Zhu, M., Rodriguez R., Wehr T., and Siebert C. (1992) Capillary electrophoresis of haemoglobins and globin chains. *J. Chromatogr.* **608,** 225–237.
22 Rodriguez, R., Zhu, M., Wehr, T., and Siebert, C. (1994) Gravity mobilization of proteins in capillary isoelectric focusing. Presented at the *Sixth International Symposium on High Performance Capillary Electrophoresis,* San Diego, CA.
23. Foret, F., Szoko, E., and Karger, B. L. (1993) Trace analysis of proteins by capillary zone electrophoresis with on-column isotachophoretic preconcentration. *Electrophoresis* **14,** 417–428.

24. Terabe, S., Otsuka, K., Ichikawa K., Tjuchiya, A., and Ando, T. (1984) Electrophoretic separations with micellar solutions and open tubular capillaries. *Anal. Chem.* **56,** 111–113.

25. Gelfi, C., Cossu, G., Carta, P., Serra, M., and Righetti, P. G. (1995) Gene dosage in capillary electrophoresis: pre-natal diagnosis of Down's syndrome. *J. Chromatogr. A* **718,** 405–412.

26. Gross, M., Gathof, B. S., Kolle, P., and Gresser, U. (1995) Capillary electrophoresis for screening of adenylosuccinate lyase deficiency. *Electrophoresis* **16,** 1927–1929.

27. Oto, M., Suehiro, T., and Yuasa, Y. (1995) Identification of mutated *p53* in cancer by non-gel-sieving capillary electrophoretic SSCP analysis. *Clin. Chem.* **41,** 1787–1788.

28. Hu, A. Z., Cruzado, I. D., Hill, J. W., McNeal, C. J., and Macfarlane, R. D. (1995) Characterization of lipoproptein a by capillary zone electrophoresis. *J. Chromatogr. A* **717(1–2),** 33–39.

29. Holmes, R. P. (1995) Measurement of urinary oxalate and citrate by capillary electrophoresis and indirect ultraviolet absorbance. *Clin. Chem.* **41,** 1297–1301.

30. Ueda, T., Maekawa, T., Sadamitsu, D., Oshita, S., Ogino, K., and Nakamura, K. (1995) Determination of nitrite and nitrate in human blood plasma by capillary zone electrophoresis. *Electrophoresis* **16(6),** 1002–1004.

31. Jariego, C. M. and Hernanz, A. (1996) Determination of organic acids by capillary electrophoresis in screening of organic acidurias. *Clin. Chem.* **42,** 477–478.

32. Marsh, D. B. and Nuttall, K. L. (1995) Methylmalonic acid in clinical urine specimens by capillary zone electrophoresis using indirect photometric detection. *J. Cap. Electrophor.* **2,** 63–67.

33. Shihabi, Z. K. and Oles, K. S. (1994) Felbamate measured in serum by two methods: HPLC and capillary electrophoresis. *Clin. Chem.* **40,** 1904–1908.

34. Abubaker, M. A., Bissell, M. G., and Petersen, J. R. (1995) Micellar electrokinetic capillary chromatography to separate steroids that are increased in congenital adrenal hyperplasia. *Clin. Chem.* **41,** 1369–1370.

35. Jenkins, M. A., Kulinskaya, E., Martin, H. D., and Guerin, M. D. (1995) Evaluation of serum protein separation by capillary electrophoresis: prospective analysis of 1000 specimens. *J. Chromatogr. B* **672,** 241–251.

36. Jenkins, M. A. and Guerin, M. D. (1995) Quantification of serum proteins using capiullary electrophoresis. *Ann. Clin. Biochem.* **32,** 493–497.

37. Jenkins, M. A. and Guerin, M. D. (1996) Optimization of serum protein separation by capillary electrophoresis. *Clin.Chem.* **42,** 1886.

38. Jenkins, M. A., O'Leary, T. D., and Guerin, M. D. (1994) Identification and quantitation of human urine proteins by capillary electrophoresis. *J. Chromatogr. B* **662,** 108–112.

39. Jenkins, M. A. (1997) Clinical application of capillary electrophoresis to unconcentrated human urine proteins. *Electrophoresis* **18,** 1842–1846.

40. Hempe, J. M. and Craver, R. D. (1994) Quantification of haemoglobin variants by capillary isoelectric focusing. *Clin. Chem.* **40,** 2288–2295.

41. Hempe, J. M., Granger, J. N., and Craver, R. D. (1997) Capillary isoelectric focusing of haemoglobin variants in the pediatric clinical laboratory. *Electrophoresis* **18,** 1785–1795.
42. Shihabi, Z. K. (1996) Analysis and general classification of serum cryoglobulins by capillary zone electrophoresis. *Electrophoresis* **17,** 1607–1612.
43. Cowdrey, G., Firth, M., and Firth, G. (1995) Separation of cerebrospinal fluid proteins using capillary electrophoresis: a potential method for the diagnosis of neurological disorders. *Electrophoresis* **16,** 1922–1926.
44. Doelman, C. J. A., Siebelder, C. W. M., Nijhof, W. A., Weykamp, C. W., Janssens, J., and Penders, T. J. (1997) Capillary electrophoresis system for haemoglobin A1c determinations evaluated. *Clin. Chem.* **43,** 644–648.

2

Serum Protein Electrophoresis

Margaret A. Jenkins

1. Introduction

Serum protein electrophoresis (SPE) is a technique that has been used in clinical laboratories for several decades to elucidate and quantitate monoclonal paraproteins. These proteins are indicative of patients with a B-cell dyscrasia, which, if untreated, could lead to the early demise of the patient.

The support media used to examine SPE have varied from the original fluid method of Tiselius (1), through paper electrophoresis, cellulose acetate (2), agarose gel (3), and high-resolution agarose gel (4). More recently, a number of scientists have used the medium of capillary electrophoresis (CE) to electrophorese human serum proteins (5–10).

Clinical laboratories in the 1990s require methods that are reliable, fast, cost-effective, and use a minimum of labor. Thus, any technique that provides automation of a previously manual technique should find acceptance within a clinical setting. The automated CE instruments now available provide the means for considerably decreasing the labor component of SPE.

The method described here was developed in a clinical laboratory over a 2-mo period. Approximately 10 buffers, usually at three different pH levels and two different ionic strengths, were tested. The results were rated, depending on whether the results produced by CE resembled the densitometer tracing of high-resolution agarose gel electrophoresis (HRAGE). Having narrowed the choice to two buffers (phosphate and boric acid), boric acid was chosen because addition of calcium lactate gave increased resolution of the β components. With this buffer, workers then set about finding the balance of the correct dilution of sample and injection time of the capillary, so that the fused-silica capillary could be calibrated and hence have a quantitative analysis for SPE.

From: *Methods in Molecular Medicine, Vol 27: Clinical Applications of Capillary Electrophoresis*
Edited by: S. M. Palfrey © Humana Press Inc., Totowa, NJ

Publication of the evaluation of the technique noted that "two cases of monoclonal IgM paraproteinaemia were detected by high resolution agarose gel electrophoresis (HRAGE) but were significantly distorted when detected by CE" *(11)*. These two samples were found in 1000 samples examined by both techniques. These aberrant samples by CE were obvious because the retention time for albumin of these samples were, in both cases, 1 min longer than the previous sample's albumin retention time. When the calcium lactate was removed from the boric acid buffer, the monoclonal bands became evident.

In 1996, using five aberrant IgM paraprotein samples and three very slow migrating monoclonal IgG samples, which also did not quantitate correctly, the 1994 published method was optimized *(12)*. It was found that increasing the pH and the ionic strength of the optimized buffer allowed correct quantitation of all of the monoclonal IgM and IgG samples. The method discussed here is the optimized method of SPE by CE.

2. Materials

2.1. Apparatus

1. An automated CE apparatus. The Applied Biosystems 270A-HT Capillary Electrophoresis System (Perkin-Elmer, Foster City, CA) is used. This instrument provides a carousel capable of handling 50 specimens, and has multiple programs that can be altered for different analytes, and a diffraction grating for precise wavelength selection. Other similar instruments may be used for SPE.
2. A suitable software system, such as Turbochrom 1V (Perkin-Elmer), should be available for analyzing the data produced by the CE electropherogram. This program allows for area under the curve to be converted to g/L for all components.
3. Calibrators: Albumin standards varying from 20–40 g/L, or four samples from patients showing minimal other pathology, and having albumin values between 20 and 40 g/L.

2.2. Capillary

1. A 72 cm × 50 μm fused silica capillary is used (Scientific Glass Engineering, Victoria, AUS). Other similar capillaries are likely to be suitable.
2. The window in the capillary is placed between 22 and 23 cm from the outlet end. A lighted match is used to burn the window, which is then wiped with methanol before placing the capillary on the instrument.
3. To bring the capillary into use, pass 1 *M* NaOH through it for 30 min, followed by 10 min with distilled water.

2.3. Stock Solutions

All solutions used for CE should be prepared volumetrically using chemicals of Analar grade. Deionized water with a resistivity greater than 10 million ohms/cm (MO/cm) is used for the preparation of all solutions. For storage conditions, see individual solutions.

1. 75 m*M* boric acid buffer, pH 10.3: Weigh out 4.635 g boric acid (BDH prod. 10058, Kilsyth, AUS). Dissolve in 950 mL distilled water. Adjust pH accurately to 10.3 with 1 *M* NaOH. Make up to 1 L. Store at room temperature for up to 3 mo.
2. 0.5 *M* Calcium lactate: Weigh out 0.15 g L(+) lactic acid (2-hydroxypropionic acid) Hemicalcium salt hydrate formula weight (FW) 109.1, Sigma L2000 (St. Louis, MO) (this allows for the 10% hydration quoted in the product). Make up to 2.5 mL with distilled water. Place in 37°C incubator for approx 20 min to allow complete solution. Mix and store at 4°C. Discard when any bacterial growth (white) is noted. Lasts approx 6 wk (*see* **Note 1**).
3. Boric acid/calcium lactate working buffer: To 50 mL of 75 m*M* boric acid buffer prepared above, add 20 μL of 0.5 *M* calcium lactate. Mix. This working buffer may be used for 2 wk (*see* **Note 2**). The working solution is left at room temperature during the day. However, it is recommended that it is stored at 4°C overnight.

3. Methods

3.1. Sample Preparation

1. Pipet 490 μL boric acid/calcium lactate into a sample cup. Pipet 10 μL serum into the buffer. Place a sample cap on the vial, and mix by inversion. Tap the bottom of the tube on the bench to remove any bubbles. Place on the carousel of the instrument.

3.2. Control Preparation

1. Choose a serum containing an IgG paraprotein of approx 20 g/L in size. Pipet 490 μL boric acid/calcium lactate into a sample cup. Pipet 10 μL control into the sample cup. Place a sample cap on the vial, and mix by inversion. Place on the carousel of the instrument.
2. Store the control serum at 4°C. Dilute freshly each day for 1 mo, then replace with a newer sample, overlapping the controls slightly.

3.3. Buffer Vials

1. Using a Sterile Acrodisc (Gelman Sciences, Ann Arbor, MI, prod. no. 4192) filter 0.1 *M* NaOH into a 4-mL buffer vial. Place white and grey tops on the buffer vial, label, and place in position 51 (*see* **Note 3**). The Acrodisc may be used for up to 3 mo if not contaminated.
2. Place distilled water into another 4-mL buffer vial, and place in position 52 (*see* **Note 3**).
3. Filter the working buffer solution through a 0.2-μm sterile Acrodisc into a buffer vial, cover with white and grey tops, and place in position 53 on the instrument.
4. The working buffer vial, if not showing any signs of contamination, may be used on the instrument for up to 2 wk. The one Acrodisc filter may be used for up to 3 mo if there are no signs of contamination.

3.4. Calibration of Serum Proteins

1. On installation of a new capillary, or on Monday of each week, choose four samples that have been analyzed for albumin on a Hitachi 911(Hoffman-La Roche, Basel, Switzerland) or similar analyzer. These samples should have albumin values of approx 20, 27, 36, and 43 g/L. If possible, choose samples with near normal pathology.
2. Add 490 µL of working buffer (boric acid/calcium lactate) to each of four labeled sample vials. Add 10 µL serum from the chosen albumin standards. Place a grey sample cap on each vial, and invert to mix. Place on the carousel of the instrument (*see* **Note 4**).
3. The area under the curve for each albumin standard is entered into the software, together with the known albumin concentration.

3.5. Electrophoresis

1. Flush the capillary for 2 min with 0.1 M NaOH, followed by water for 1 min and electrophoresis buffer for 2 min.
2. Set the wavelength to 200 nm, applied voltage to 20 kV, and the run time to 12 min (*see* **Note 5**).
3. Load the sample for 2 s, using a vacuum set to 5 in.

3.6. Processing Calibration Data

1. Record the area under the curve for each albumin peak and albumin concentration.
2. Calibration type: Use a curve fit.

3.7. To Cut Electropherogram at Preferred Place for Peak Measurement

1. With the Turbochrom software, this is done through Reprocess.
2. Select the electropherogram required.
3. Process, baseline events, Start New Peak Now, click on Start New Peak Now on valley at either side of peak. Reprocess. Return.
4. Display peak report: this will give quantitation of monoclonal band that has been cut.

3.8. Analysis of Electrophoretic Patterns

1. To assist with the interpretation of an electrophoretic pattern, the chemical quantitation of total protein and albumin, and the patient's history, are printed automatically on each worksheet.
2. Reports give an overall assessment of components of the electropherogram that are elevated or decreased, indicating the severity of any increase or decrease as mild, moderate, or marked.
3. The only quantitation figures reported are for any monoclonal band or bands.
4. **Figures 1A–D** show a normal serum electropherogram, a monoclonal band of 18 g/L, an acute phase response indicated by moderately increased α-1 and 2, mildly

increased C3 and a possible elevation of CRP in the γ area, and a patient with a double free κ light chain band with associated decreased residual γ-globulins.
5. **Figures 2** and **3** illustrate poor quality electropherograms, possible causes are low lamp energy (*see* **Note 6**), protein buildup, dirty buffer vials, or jagged capillary end (*see* **Notes 7–10**).

4. Notes

1. Discard the calcium lactate when any white bacterial growth is seen. This often occurs about 6 wk after it is made. Use of the calcium lactate at this stage can cause spikes.
2. When a fresh batch of working buffer is made up for dilution of specimens, do not forget to change the running buffer vial at position 53 of the instrument. Otherwise, you may get spikes in the gamma region of the electropherogram, which are indicative of slight variations in buffer.
3. Replace the 0.1 M sodium hydroxide and water vials in position 51 and 52 at least twice a week.
4. Since quantitative values from the CE are being reported, it is essential that the pipets used for dilution of the sample in buffer are clean, correctly calibrated, and well maintained. Calibration of pipets should be routinely checked using dye dilution/spectrophotometry or weighing techniques every 6 mo.
5. If any protein appears after the albumin peak, wash the capillary in 1 M NaOH for 5 min, then wash with water, followed by a rerun of the sample. Occasionally, there is buildup of protein on the capillary.
6. If the baseline is noisy, i.e., there is visible wobbling in the baseline and it is not a perfectly straight line, check the absorbance of the sample and reference at 238 nm through the Service menu, Self Tests and Detector. The absorbance at 238 nm, according to the manufacturers, should be greater than 0.25. The baseline will show noise when the absorbance is about 0.19. The lamp will definitely need changing at 0.18. An example of a sample with a noisy baseline is shown in **Fig. 2**.
7. CE is a very sensitive technique; hence, any contaminant is likely to show up as a small peak. The author has found that the washing of the buffer vials is best done by the people operating the CE instrument. The routine is as follows. Place distilled water into a small plastic container. Any used buffer vials taken off the CE instrument have their contents discarded, and the buffer vials are placed in the plastic container of distilled water. Also, the grey tops from the samples and buffer vials are reused. The sample tubes are discarded. The grey tops are placed in the distilled water to soak. Approximately once a fortnight, rinse the contents of the plastic tub in more distilled water, rub any marks off the sides of the buffer vial tubes, and place the tubes and tops on low lint tissues in a large weighing tray. This tray is placed in an oven at 70°C for 2–3 h. Do not bake the tops.
8. If the amplitude of the protein peaks becomes small, it may be because the inside of the inlet of the capillary has a buildup that is not letting the correct amount of sample be aspirated. This situation can be remedied by carefully cutting 0.5 cm from the end of the capillary. If the capillary has just been installed, another

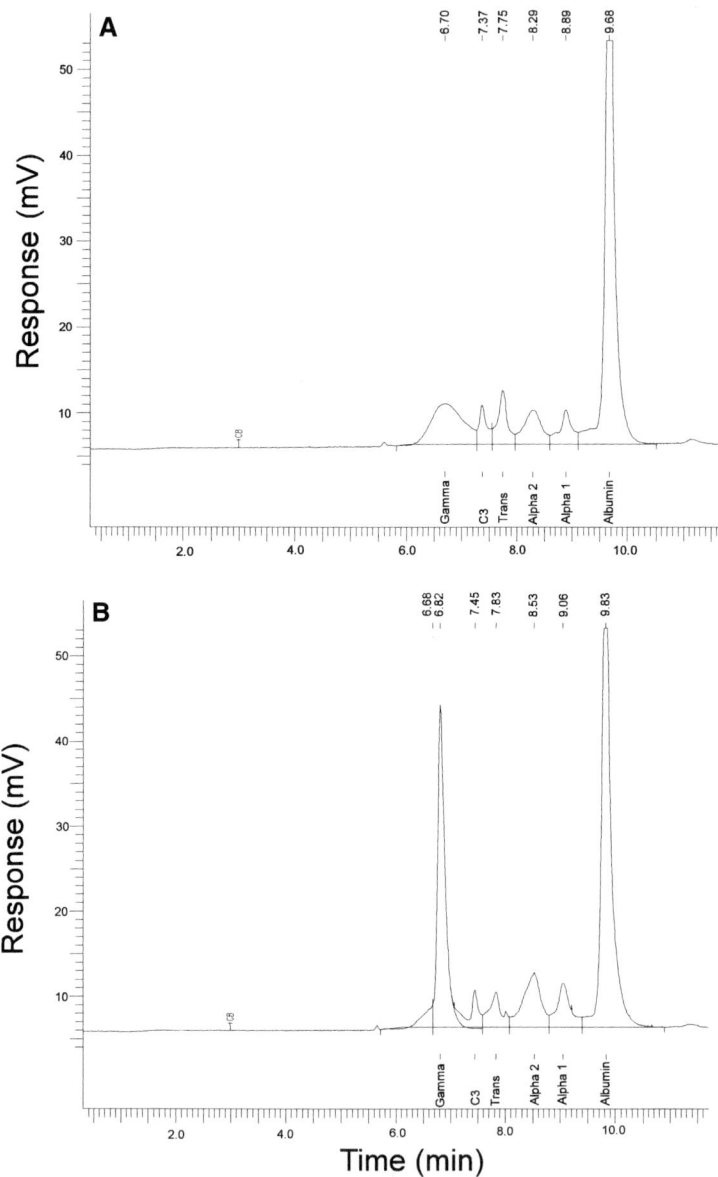

Fig. 1. Capillary electropherograms showing (**A**) normal serum electrophoresis, (**B**) IgG (k) monoclonal band 18 g/L with moderate associated immune paresis.

alternative for small peak height is that the capillary window is not correctly seated, i.e., the polyimide cover of the capillary is covering half of the window. This situation may be corrected by reseating of the capillary window.

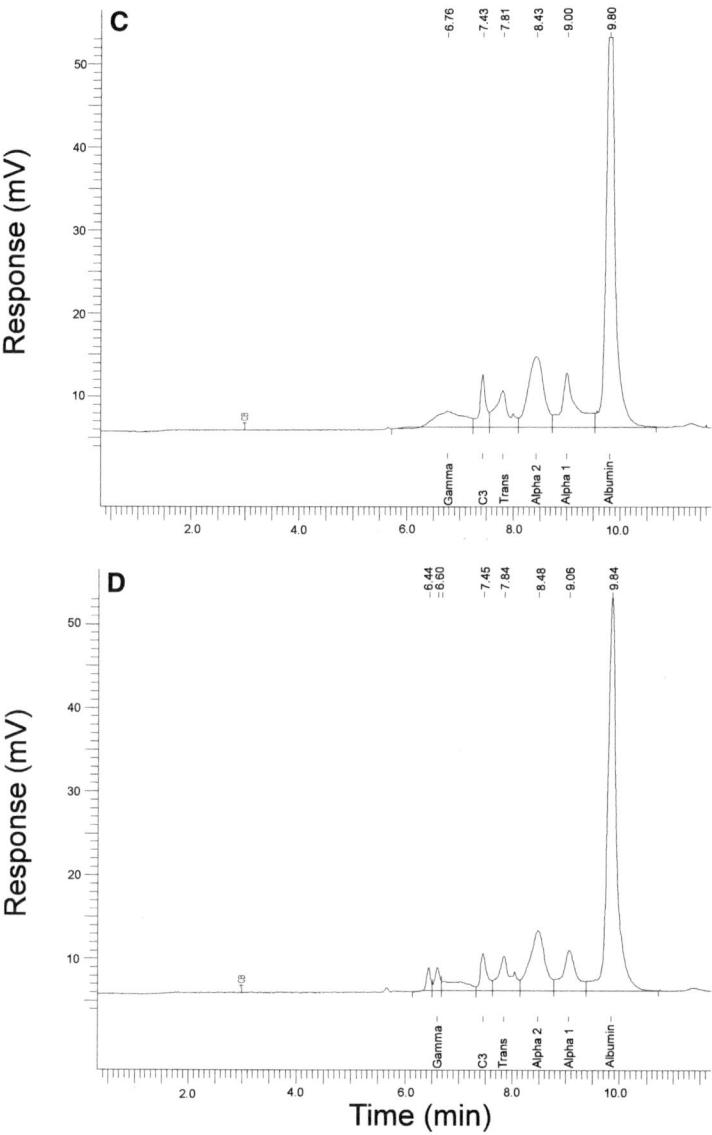

(**C**) Increased acute-phase reactants with a probable increased CRP in mid-γ, and (**D**) double free κ light-chain band with associated decreased residual γ-globulins. Electrophoretic conditions as described in **Subheading 3.5.** of Methods.

9. If the inlet end of the capillary is cut after installation of the capillary, the fused silica coating may be jagged and slowly release particles into the buffer, which is subsequently aspirated. These particles may show up on the electropherogram as spikes (*see* **Fig. 3**).

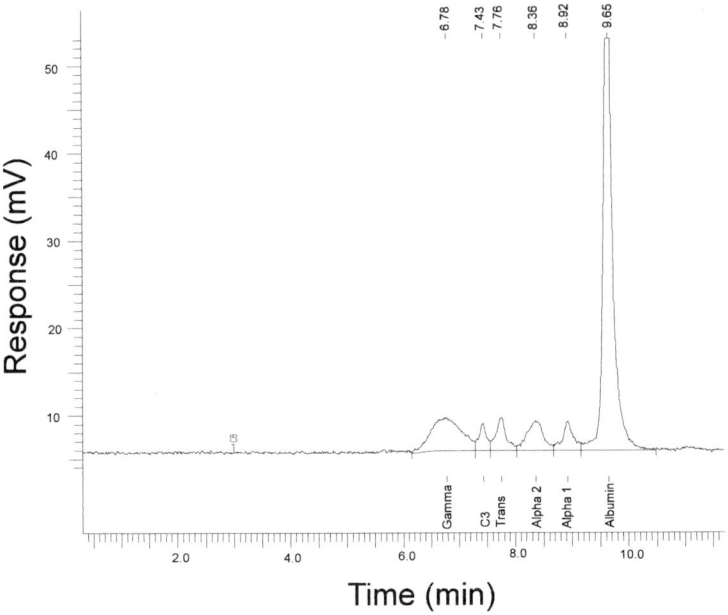

Fig. 2. Electropherogram showing a noisy baseline caused by decreased energy of deuterium lamp. For comparison, *see* **Fig. 1A,** which has a normal baseline.

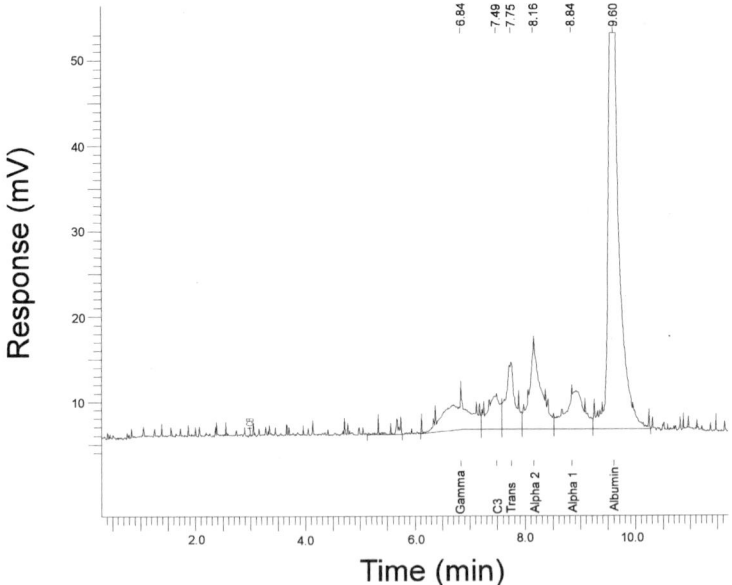

Fig. 3. Electropherogram of sample run after fused silica capillary has been scored and cut with a capillary cutter. Spikes caused by release of fused silica from inside of capillary.

10. When an electropherogram is a straight line, check for capillary integrity. This is done by flushing air through the capillary from an empty buffer space. If bubbles are seen coming through the outlet, then the capillary is not blocked. If there are no bubbles, try 5 min with 1 *M* NaOH to try to unblock the capillary. It is also worth checking that the inlet end of the capillary on the Applied Biosystems CE system is parallel to the anode (RH end of the capillary). If the capillary has hit a buffer tube, it may be at 45 degrees to the electrode, and not aspirating as the program indicates. This problem will not happen with cassette-type CE instruments. Another aspect to check is to redilute the specimen, and check that there is actually sample in the sample cup.

References

1. Tiselius, A. (1937) New apparatus for electrophoretic analysis of colloidal mixtures. *Trans. Faraday Soc.* **33,** 524–531.
2. Riches, P. G. and Kohn, J. (1987) Improved resolution of cellulose acetate membrane electrophoresis. *J. Ann. Clin. Biochem.* **24,** 77–79.
3. Jeppsson, J.-O., Laurrell, C.-B., and Franzen, B. (1979) Agarose gel electrophoresis. *Clin. Chem.* **25,** 629–638.
4. Johanssen, B. G. (1972) Agarose gel electrophoresis. *Scand. J. Clin. Lab. Invest.* **29 (Suppl 124),** 7–19.
5. Gordon, M. J., Lee, K-J., Arias, A. A., and Zare, R. N. (1991) Protocol for resolving protein mixtures in capillary zone electrophoresis. *Anal. Chem.* **63,** 69–72.
6. Chen, F.-T. A., Liu, C.-M., Hsieh, Y.-Z., and Sternberg, J. C. (1991) Capillary electrophoresis—a new clinical tool. *Clin. Chem.* **37,** 14–19.
7. Kim, J. W., Park, J. H., Park, J. W., Doh, H. J., Heo, G. S., and Lee, K.-J. (1993) Quantitative analysis of serum proteins separated by capillary electrophoresis. *Clin. Chem.* **39,**689–692.
8. Jenkins, M. A. and Guerin, M. D. (1995) Quantification of serum proteins using capillary electrophoresis. *Ann. Clin. Biochem.* **32,**493–497.
9. Dolnik, V. (1995) Capillary zone electrophoresis of serum proteins: study of separation variables. *J. Chromatogr. A* **709,** 99–110.
10. Lehmann, R., Liebich, H. M., and Voelter, W. (1996) Application of capillary electrophoresis in clinical chemistry: developments from preliminary trials to routine analysis. *J. Capillary Electrophoresis* **3,** 89–110.
11. Jenkins, M. A., Kulinskaya, E., Martin, H. D., and Guerin, M. D. (1995) Evaluation of serum protein separation by capillary electrophoresis: prospective analysis of 1000 specimens. *J. Chromatogr. B.* **672,** 241–251.
12. Jenkins, M. A. and Guerin, M. D. (1996) Optimization of serum protein separation by capillary electrophoresis. *Clin. Chem.* **42,** 1886.

3

Urine Proteins

Margaret A. Jenkins

1. Introduction

Urine protein electrophoresis has been used primarily to confirm the presence or absence of Bence Jones protein, a small mol-wt protein consisting of either monoclonal free κ or free λ light chains. Bence Jones protein is significant in multiple myeloma patients, because nephropathy can develop in 70% of patients exhibiting Bence Jones protein *(1–3)*.

Normal urinary protein excretion is less than 0.15 g/d, and the major component is albumin. In renal disease, proteinuria may be classified as either glomerular or tubular. Glomerular proteinuria, which can be associated with infections, neoplasia, some hereditary diseases, and certain drug exposure, is characterized by the loss of protein of mol wt of albumin or greater. Tubular proteinuria is caused by a decreased capacity of the tubules to reabsorb proteins of small mol wt, such as β-2-microglobulin or α-2-microglobulin. Tubular proteinuria can be caused by chronic exposure to metals such as cadmium dust, lead, mercury, or gold, and can also be seen in pyelonephritis, renal transplant rejection, Fanconi's syndrome, or sarcoidosis.

Urine protein electrophoresis of concentrated urine specimens has previously used support media similar to serum protein electrophoresis. These support media included paper, cellulose acetate, agarose, and high-resolution agarose gel. The extreme sensitivity of CE made early attempts to use the CE technique for urine electrophoresis difficult, because of the large number of peaks found. These peaks were assumed to be small molecules and breakdown products, probably peptides *(4)*.

Originally, this laboratory used three methods to examine concentrated urine specimens by CE. These included anion-exchange resin treatment of the urine to remove nonprotein components, the use of abnormal urine

From: *Methods in Molecular Medicine, Vol 27: Clinical Applications of Capillary Electrophoresis*
Edited by: S. M. Palfrey © Humana Press Inc., Totowa, NJ

containing previously identified proteins, and the addition of known analytes to urine specimens, such as albumin, phosphate, nitrate, and oxalate. Using these techniques, albumin and Bence Jones protein were identified, and a correlation of 71 concentrated urine specimens for albumin and Bence Jones protein were subsequently published, using CE and commercial high resolution agarose gel electrophoresis *(5)*.

For the past 20 yr, hospital scientists performing urine protein electrophoresis have begun by concentrating the urine, usually using commercial urine concentrators. These commercial urine concentrators have become increasingly expensive in the last 3 yr. Also, at least 30 min was required for the concentration of a urine specimen. Hence, in 1996, recognizing the extreme sensitivity of CE , the use of spun, unconcentrated urine for urine protein electrophoresis was investigated *(6)*. By manipulating the dilution of the spun urine with running buffer, results virtually identical to those obtained previously with concentrated urine specimens were obtained. Workers studied 22 urine specimens using unconcentrated vs concentrated urine electrophoresis by CE for both Bence Jones protein and albumin, and found a correlation of 0.956 and 0.996, respectively *(7)*.

This chapter on Bence Jones protein uses spun, unconcentrated urine specimens for the CE of urinary proteins.

2. Materials

2.1. Apparatus

1. An automated CE apparatus, such as an Applied Biosystems 270A-HT Capillary Electrophoresis System (Perkin-Elmer, Foster City, CA), is used. This instrument provides a carousel capable of handling 50 samples, and has multiple programs able to be adapted to different analytes, and a diffraction grating for wavelength selection. Other similar instruments may be used for urine protein electrophoresis.
2. A 72 cm × 50 μm fused-silica capillary is used (Scientific Glass Engineering, Victoria, AUS). Other similar capillaries are likely to be suitable.
3. A software system, such as Turbochrom IV (Perkin-Elmer), should be optimally available for analyzing the data produced by the CE electropherogram. However, any software program that records the area of all the individual peaks will be sufficient. If peaks are not cut to the operator's satisfaction, then the software should have the ability to cut individual peaks.

2.2. Stock Solutions

All solutions used for CE should be prepared volumetrically, using chemicals of Analar grade. Deionized water, with a resistivity greater than 10 MO/cm, was used for the preparation of all solutions. For storage conditions, see individual solutions.

1. 150 mM boric acid buffer, pH 9.7: Weigh out 4.635 g boric acid (BDH, Kilsyth, AUS). Dissolve in 450 mL distilled water. Adjust pH accurately to 9.7 with 1 M NaOH, and the volume to 500 mL. Store at room temperature for up to 18 mo.
2. 0.5 M Calcium lactate: Weigh out 0.15 g L(+) lactic acid (2-hydroxypropionic acid) hemicalcium salt hydrate formula weight (FW) 109.1, Sigma L2000 (St. Louis, MO) (This allows for the 10% hydration quoted on the product). Make up to 2.5 mL with distilled water. Place in a 37°C incubator for approx 20 min to allow complete dissolution. Mix. Store at 4°C. Discard when any bacterial growth (white) is noted. Lasts approx 6 wk.
3. Boric acid/calcium lactate working buffer: To 50 mL 150 mM boric acid, add 0.1 mL 0.5 M calcium lactate. Mix. The working boric acid/calcium lactate reagent is stored at 4°C overnight when not in use.

3. Methods

3.1. Sample Preparation

1. Label an Eppendorf tube with the patient details. Spin the Eppendorf tube of urine at 1200g for 5 min (*see* **Notes 1** and **2**).
2. Pipet 60 µL working reagent (boric acid/calcium lactate) into a sample cup. Add 40 µL spun urine specimen, and mix. Cap the sample tube, and place on the carousel.

3.2. Buffer Vials

1. Filter 0.1 M NaOH through a 0.2 µm filter (Sterile Acrodisc, prod. no. 4192, Gelman Sciences, Ann Arbor, MI) into a 4-mL vial. Label and place in position 51. The Acrodisc may be used for up to 3 mo if not contaminated.
2. Place distilled water into another 4-mL buffer vial, and place in position 52.
3. Filter the working reagent (boric acid/calcium lactate) through another 0.2-µm sterile filter into a 4-mL reagent vial for use; place in position 55. The working buffer vial may be used in the instrument for up to 2 wk provided it does not show any sign of contamination (*see* **Note 3**).

3.3. Electrophoresis

1. Wavelength 200 nm, temperature 30°C, analysis time 15 min at 18 kV.
2. Flush the capillary for 2 min with 0.1 M NaOH, followed by 1 min with water and 2 min with run buffer.
3. Load the sample for 5 s using a vacuum of 5 in.

3.4. Calculation of Protein in Urine Sample

1. Use a manual trichloracetic acid method using a known albumin standard and a recognized QC material, such as Bio-Rad Lyphochek (Hercules, CA) for the accurate quantitation of urine total protein (*see* **Note 4**).
2. The first peak seen by this method has been proved by two independent research groups to be a combination of urea and creatinine *(5,8)*. Bence Jones peaks may

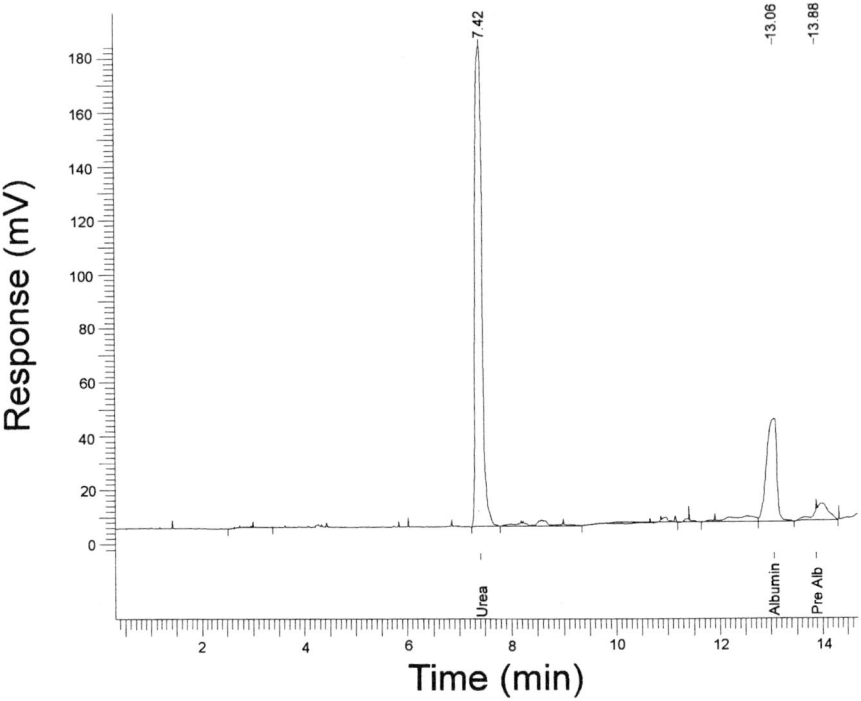

Fig. 1. Capillary electropherogram of an unconcentrated normal urine specimen. Urine protein 0.09 g/L. Electrophoretic conditions: 150 mM boric acid (+ Ca lactate), pH 9.7; injection 5 s, 1.27×10^2 mm vacuum injection; voltage 18 kV; measurement 200 nm.

occur from the urea/creatinine peak to the α-2 region. However, they are usually cathodic to transferrin. Ions such as phosphate, nitrate, and oxalate are found anodic to the prealbumin peak, and should be disregarded for quantitation purposes (*see* **Notes 5** and **6**).

3. The current method of calculating the percentage of Bence Jones protein in the specimen is to manually add all the protein peak areas, and then calculate the proportion of the Bence Jones peak relative to all the protein peaks.
4. This method is also used to calculate the percentage of albumin in the sample.

3.5. Analysis of Electrophoretic Patterns

1. Report the total protein of the urine specimen, and whether there is any Bence Jones protein present.
2. If the total protein is greater than 0.2 g/L, also indicate whether the proteinuria is glomerular or tubular in origin, or if it is a mixed glomerular/tubular proteinuria.
3. An example of a normal urine is shown in **Fig. 1**, with two specimens containing Bence Jones protein shown in **Figs. 2** and **3** (*see* **Note 7**).

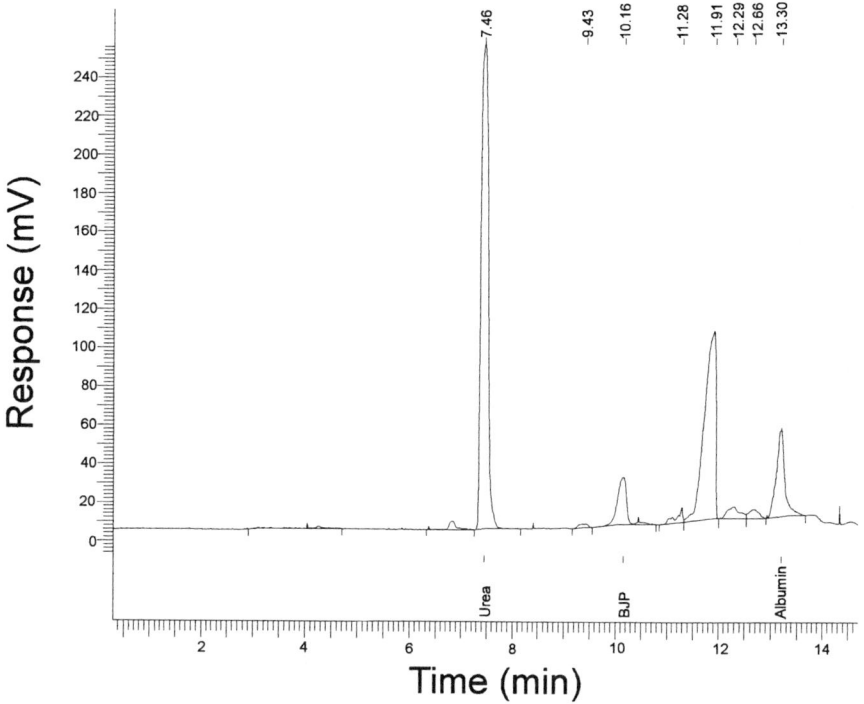

Fig. 2. Capillary electropherogram of a patient with a urine protein of 1.03 g/L, who has a Bence Jones protein concentration of 0.13 g/L, which, by immunofixation, was shown to be free λ light chains. Electrophoretic conditions as in **Fig. 1.**

4. Occasionally intact immunoglobulin as well as Bence Jones protein, may be found in the urine. Such a specimen is shown in **Fig. 4** (*see* **Note 8**).
5. Other illustrations of various urine patterns may be found in ref. *6.*

4. Notes

1. Urine specimens must be spun before being diluted in buffer, because of the particulate matter that is often found in urine. This is very obvious when some urine specimens have been refrigerated overnight.
2. The value to all laboratories of using unconcentrated urine specimens, instead of concentrated urine specimens for analysis, relates to the cost-saving of the concentrator (>$5 per concentrator), and the time saved (approx 30 min) by not concentrating the urine specimen.
3. The urine buffer vial on the instrument in position 55 is able to be used for 2 wk, providing there is no buffer depletion. The 0.1 *M* NaOH and water vials are changed twice a week.
4. Several automated urine protein methods, such as Coomassie or benzethonium chloride, may underestimate the total protein if Bence Jones protein is present.

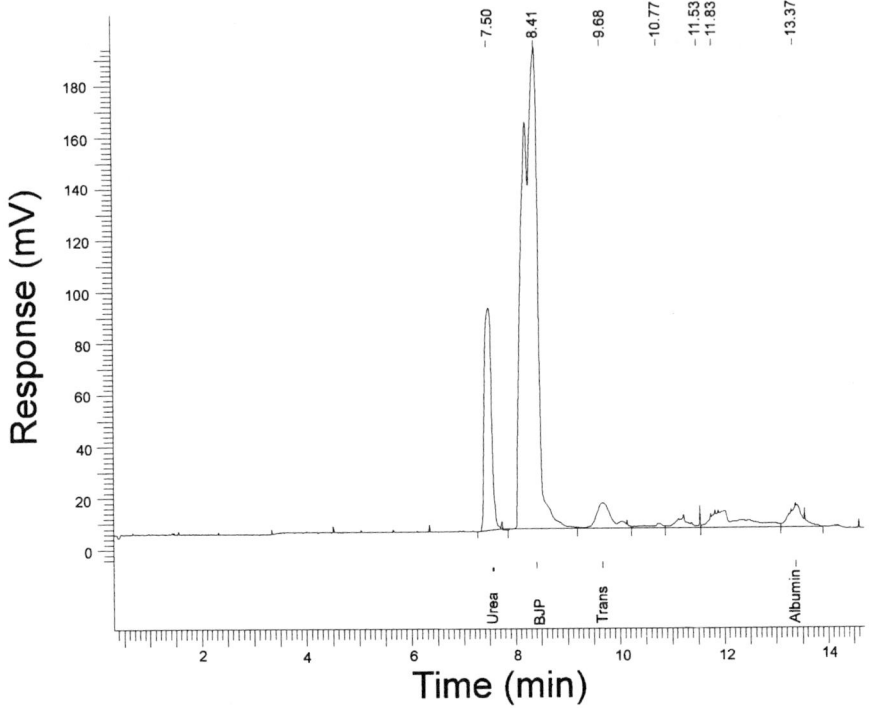

Fig. 3. Capillary electropherogram of urine with a total protein of 8.51 g/L. The double banded Bence Jones band which quantitated at 7.0 g/L, was shown by immunofixation to be caused by free κ light chains. Electrophoretic conditions as in **Fig. 1.**

However, if Bence Jones is not present, these automated methods are suitable for quantitation of urine total protein.

5. In the setting up of the calculation software, it is optimal for the highest measured peak in the urine chromatogram to reach the top of the page, the scaling on the LHS of the electropherogram is adjusted to this highest peak. In urine, this highest peak is often the urea/creatinine peak (*see* **Fig. 1**). However, with large amounts of Bence Jones protein it may be the light chain that is the highest peak (*see* **Fig. 3**).

6. The ratio of fronts (Rf) values in paper chromatography relate to the distance a substance such as an amino acid has moved, compared to the distance the solvent has moved under stated experimental conditions. Rf values are not used routinely in CE. However, in urine electropherograms, the time (in min) for the appearance of the albumin peak divided by the time (in min) for the urea/creatinine peak appears to be a constant of 1.78 ± 0.02.

7. If a small peak is found in the region between the urea/creatinine peak and α-2, the most reliable way to exclude Bence Jones protein is to apply 2 μL of uncon-

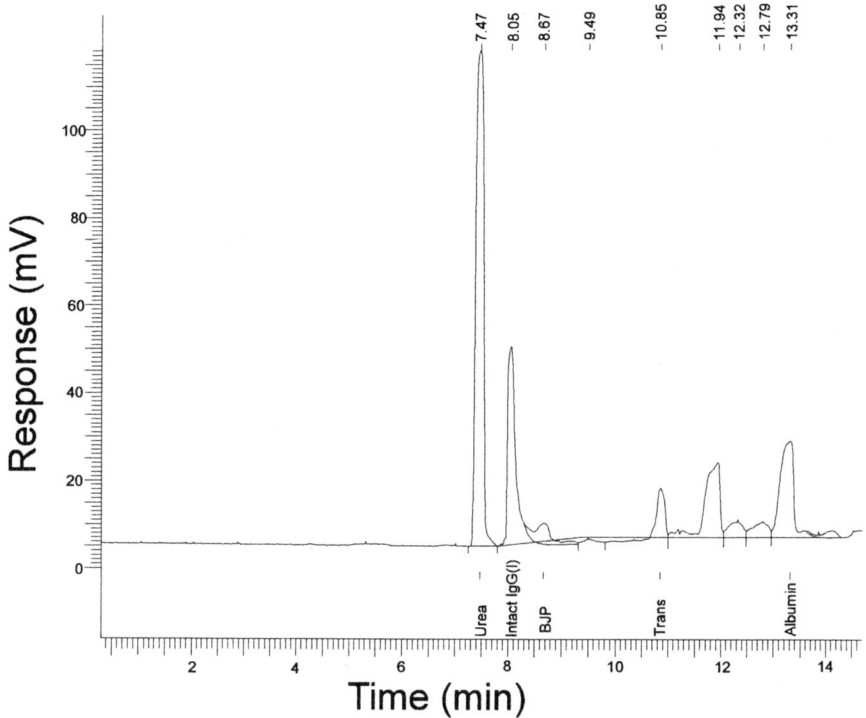

Fig. 4. Electropherogram of unconcentrated urine from a patient with a large mono-clonal IgG (λ) band in the serum. Urine shows both intact IgG (λ), as well as free λ light chains. Electrophoretic conditions as in **Fig. 1.**

centrated urine to two tracks on either isoelectric focusing (IEF) or electrophoretic (EP) gel, and to immunofix for free κ and λ light chains silenus (Amrad Operations Pty. LTD, Melbourne, AUS) when the IEF or EP gel is complete.

8. If the urine of a patient with a serum monoclonal band is being examined, and a peak is seen between the urea/creatinine and α_2, it is preferable to apply 2 μL of unconcentrated urine to either an IEF or EP gel, and to immunofix for the heavy and light chain of the band when the IEF or EP is complete. In rare cases, the band in the urine may be caused by the opposite light chain, but this is highly unusual.

References

1. Rota, S., Mougenot, B., Baudouin, B., De Meyer-Brasseur, M., Lemaitre, V., Michel, C., et al. (1987) Multiple myeloma and severe renal failure: A clinico-pathologic study of outcome and prognosis in 34 patients. *Medicine* (Baltimore) **66,** 126–137.
2. Johnson, W. J., Kyle, R. A., and Pineda, A. A. (1990) Treatment of renal failure associated with multiple myeloma: plasmaphoresis, hemodialysis, and chemo-therapy. *Arch. Intern. Med.* **150,** 863–869.

3. Alexanian, R., Barlogie, B., and Dixon, D. (1990) Renal failure in multiple myeloma: pathogenesis and prognostic implications. *Arch. Intern. Med.* **150,** 1693–1695.
4. Chen, F.-T. A., Liu, C.-M., Hsieh, Y.-Z., and Sternberg, J. C. (1991) Capillary electrophoresis: a new clinical tool. *Clin. Chem.* **37,** 14–19.
5. Jenkins, M. A., O'Leary, T. D., and Guerin, M. D. (1994) Identification and quantitation of human urine proteins by capillary electrophoresis. *J. Chromatogr. B.* **662,** 108–112.
6. Jenkins, M. A. (1997) Clinical application of capillary electrophoresis to unconcentrated human urine proteins. *Electrophoresis* **18,** 1842–1846.
7. Jenkins, M. A. (1998) Capillary Electrophoresis in the clinical laboratory. *Today's Life Sci.* **10(2),** 22–26.
8. Guzman, N. A., Berck, C. M., Hernandez L., and Advis, J. P. (1990) Capillary electrophoresis as a diagnostic tool: determination of biological constituents present in urine of normal and pathological individuals. *J. Liq. Chromatogr.* **13,** 3833–3848.

4

Electrophoresis of Cerebrospinal Fluid

Geoffrey Cowdrey, Maria Firth, and Gary Firth

1. Introduction

Under normal circumstances, cerebrospinal fluid (CSF) is a clear and color-less fluid that is formed in the ventricles of the brain. It is in close proximity to the surface of both the brain and spinal cord, and, as a result, the analysis of CSF proteins and other constituents in samples taken by lumbar puncture have long been used as an aid in the diagnosis of neurological disorders. Various electrophoretic methods have been used, including agar gel *(1)*, polyacryla-mide *(2)*, two-dimensional *(3)*, and isoelectric focusing *(4)*, with the aim of detecting profiles that are diagnostic, especially in the case of proteins. These techniques have been labor-intensive, time-consuming, and, at best, only semiquantitative. This chapter describes how the technique of capillary elec-trophoresis (CE) in free solution (FSCE) can be used to provide a very fast, sensitive, and reproducible method for the analysis of CSF constituents, using only nanoliter volumes of sample *(5)*. Furthermore, on line detection of the separated constituents, using UV absorption, allows accurate quantitation.

2. Materials

1. All separations are carried out using an Applied Biosystems 270A capillary elec-trophoresis apparatus (Perkin-Elmer, Warrington, UK.) Data is collected using a Hewlett Packard HP 3394 integrator (Bracknell, UK).
2. The CE apparatus is fitted with a hydrophilic coated fused-glass capillary (CElect-P150, Supelco, Dorset, UK). The total length is 55 cm, with id 50 μm. The distance between the capillary detection window and the sample inlet is 35 cm.
3. Methyl cellulose, viscosity 1500 centipoises at 2% (Sigma, Dorset, UK).
4. Electrophoresis buffer: 40 mM borate, containing 0.4 g/L methyl cellulose, pH 10. Weigh out 15.25 g sodium tetraborate, and dissolve in about 900 mL distilled water. Adjust the pH to 10 with the addition of 1 mM-sodium hydroxide solution,

From: *Methods in Molecular Medicine, Vol 27: Clinical Applications of Capillary Electrophoresis*
Edited by: S. M. Palfrey © Humana Press Inc., Totowa, NJ

and then make up to 1 L with distilled water. Heat 100 mL borate buffer to about 60°C and add 0.4 g methyl cellulose powder, with constant stirring. Cool to room temperature under running water with constant shaking. Place at 4°C overnight, to allow complete hydration of the methyl cellulose. Add the clear methyl cellulose solution to the rest of the borate buffer, to bring the volume back to 1 L. Mix well and store at 4°C. The buffer is stable for at least 2 mo (*see* **Notes 1** and **2**). Before use, filter sufficient buffer through a 0.45-μm disposable filter, and then degas by placing in a vacuum for a few minutes. The CE buffer is then ready for use.

3. Methods
3.1. Sample Preparation

CSF samples are used undiluted, unless the total protein concentration exceeds 600 mg/ L, in which case they should be diluted in the CE buffer that has previously been diluted 1:10 with distilled water. Samples may be stored undiluted and without preservative for up to 4 wk at 4°C.

3.2. Parameter Settings for CE Apparatus

1. Temperature 30°C.
2. Sample injection 4 s, hydrodynamically (*see* **Note 3**).
3. Wavelength 200 or 214 nm (*see* **Notes 4** and **5**).
4. Separation voltage 25 kV.
5. Analytical time 35 min (*see* **Note 6**).

3.3. Cycle for Running the Method

1. Rinse the capillary with 0.1 m*M* NaOH for 1 min. This rinse solution is placed in position 1 on the carousel, and removes any sample remaining from the previous run.
2. Rinse the capillary with electrophoresis buffer for 4 min.
3. Inject the sample, e.g., 4 s.
4. Separation voltage 25 kV.
5. Stop run after 35 min. No more peaks are detected after this time.
6. Rinse capillary in 0.1 mol NaOH for 1 min.
7. Rinse the capillary in 40 m*M* borate buffer for 4 min.
8. The capillary is now ready for the next sample. If the capillary is not going to be used again immediately, rinse with distilled water and leave. For longer-term storage, the capillary is best flushed out with nitrogen to dry.

3.4. Interpretation of CSF Capillary Electrophoresis Patterns

1. A typical run allows between 20 and 25 peaks to be separated, which includes proteins and other CSF components, when a wavelength of 200 nm is selected (**Fig. 1**). Most CSF samples analyzed by this technique show patterns that appear similar to that shown in Fig. 1, although there are often minor differences both quantitatively and qualitatively.

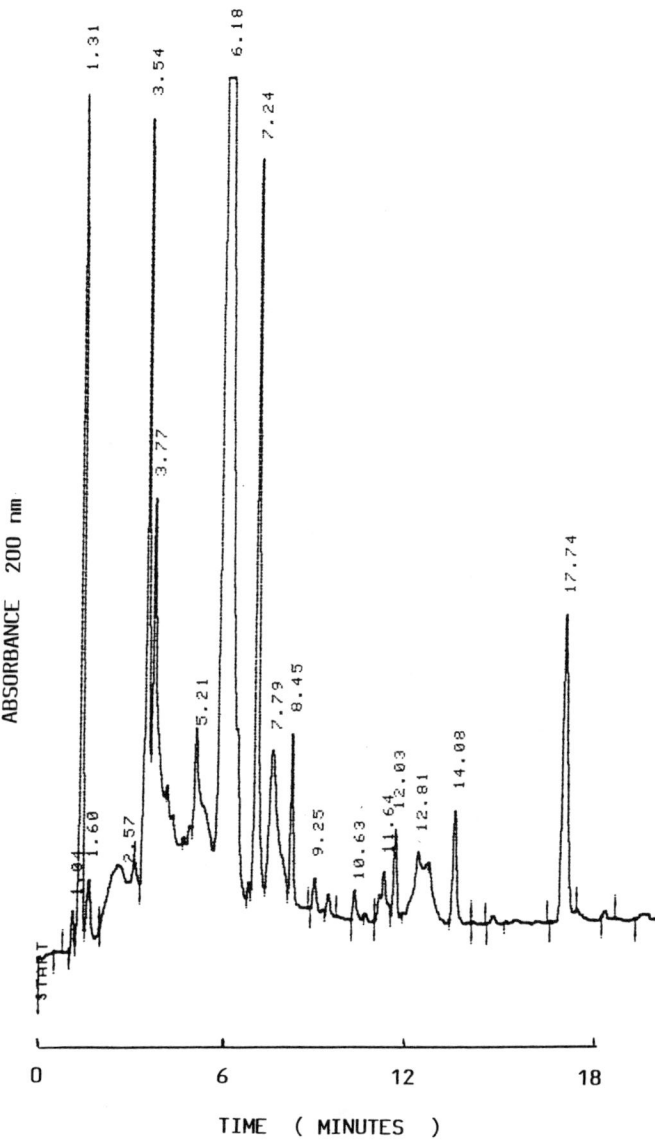

Fig. 1. Separation of CSF proteins showing a typical normal pattern. Separation buffer 40 m*M* borate, pH 10.0, containing 0.4 g/L methyl cellulose, hydrophilic-coated capillary 50 µ id × 55 cm (35 cm to detector), voltage 25 kV, temperature 30°C, 4 s injection.

2. There are, however, instances when CSF samples give patterns in which there are clear differences from the normal ones. Examples of these are shown in **Figs. 2–4**.
3. When the pattern obtained from CSF (**Fig. 5A**) is compared with the corresponding serum sample (**Fig. 5B**), the most obvious differences between the

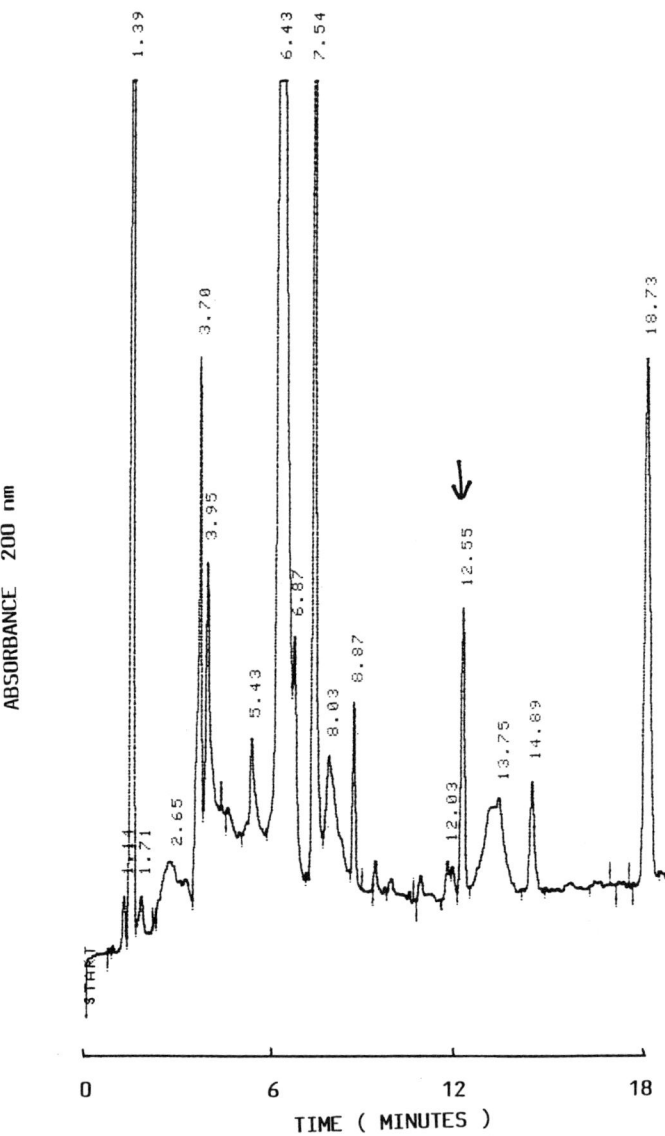

Fig. 2. Separation of CSF proteins, showing an increased concentration of acidic protein peak (arrow). Conditions as in **Fig. 1**.

two patterns are the prominent peak A in CSF, which is absent from serum; the relatively broad peak B in serum (probably IgG), but very small counterpart in CSF; the prominent split peak C in CSF, the leading part of which is absent from serum (probably the τ protein or desialayted transferrin); and the striking

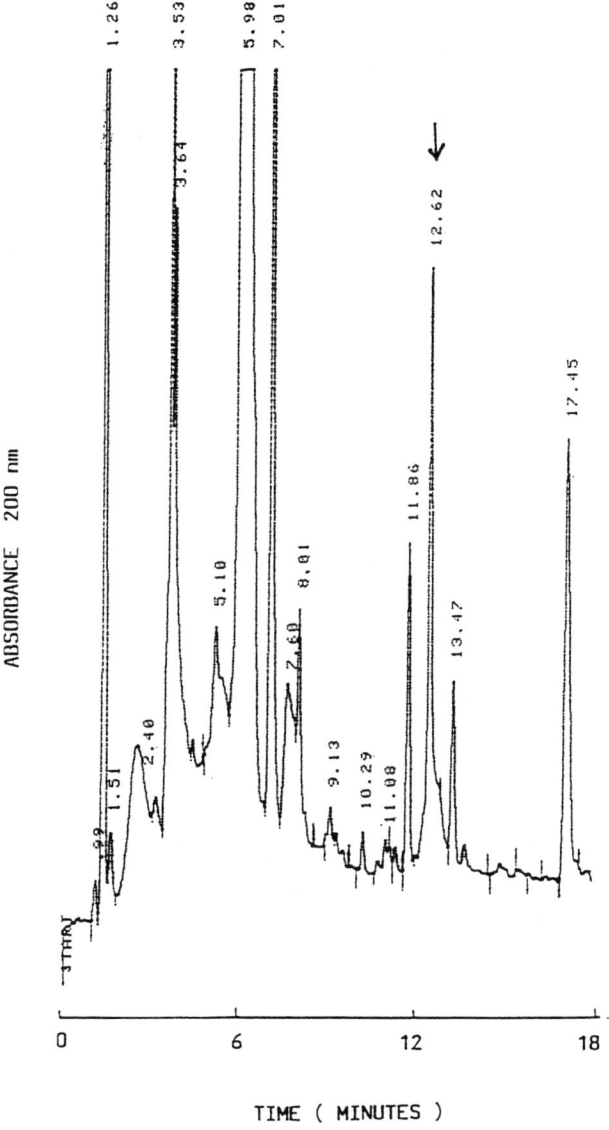

Fig. 3. Separation of CSF proteins showing marked increase in acidic protein peak (arrow). Conditions as in **Fig. 1**.

number of peaks with long migration times in region D to E but which are absent from serum.

4. In FSCE, using a borate buffer, pH 10, the electroendosmotic flow is substantial, and, under normal polarity, is toward the cathode. The elution order of peaks is,

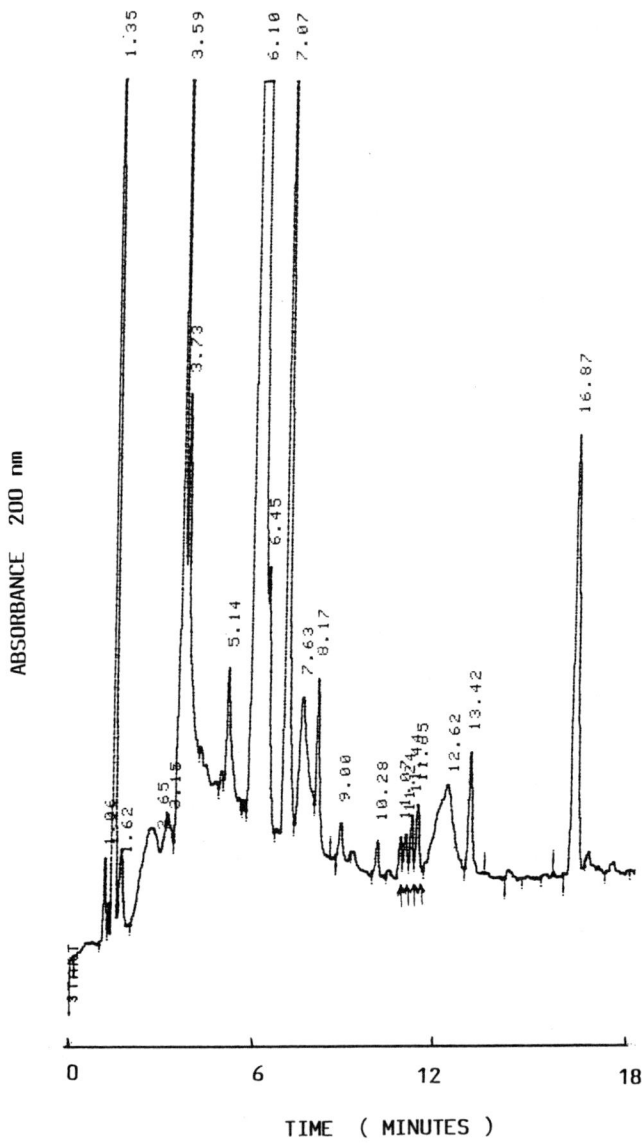

Fig. 4. Separation of CSF proteins showing a group of abnormal protein peaks (arrow). Conditions as in **Fig. 1**.

therefore, first cations, then neutrals, and finally anions; the latter also migrate toward the cathode, because the electroendosmotic flow exceeds electrophoretic migration. The migration order in FSCE depends on the charge:size ratio of the analyte, and so it is likely that the peaks with long migration times in CSF are

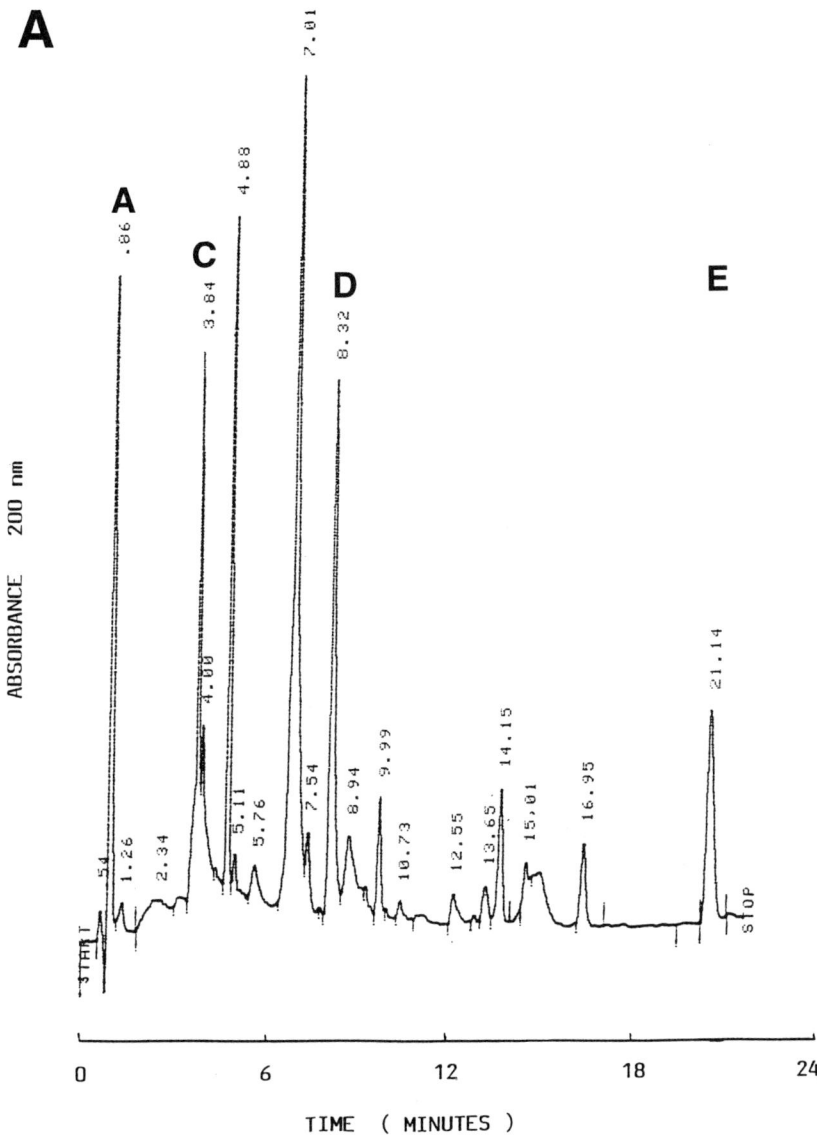

Fig. 5. Separation of CSF (**A**) and serum proteins (**B**), from same patient, showing differences between patterns. Protein peak A is only present in CSF, B is present in a much smaller amount in CSF, C is a split peak in CSF, but a single peak in serum, region D to E contains many acidic proteins in CSF that are absent from serum. Conditions as in **Fig. 1,** except voltage is 20 kV.

Fig. 5B *(shown on next page).* Relatively acidic constituents. The appearance of so many relatively acidic peaks with long migration times is somewhat surpris-

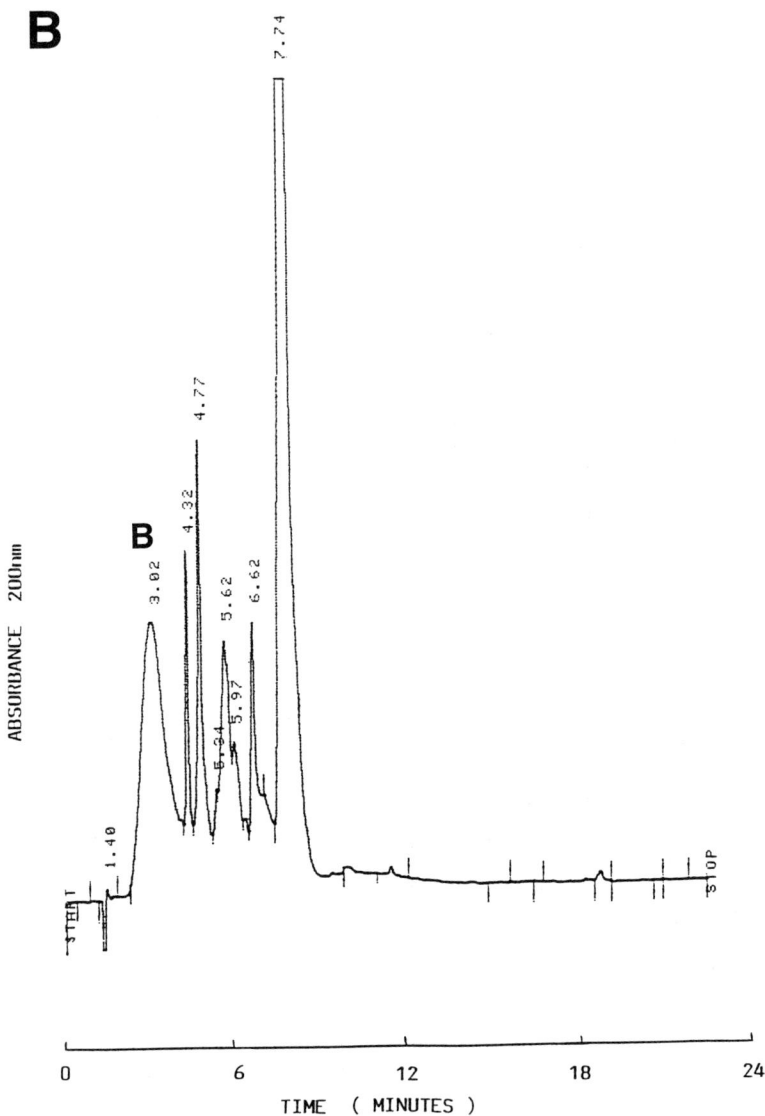

ing. These peaks appear to be CSF-specific, since they are not detected in serum even at only 1:5 dilution. The significance of these peaks in CSF is not clear.

4. Notes

1. When preparing the 40 mM borate buffer, it is best to make up a relatively large volume, e.g., 1 L, so as to minimize any batch-to-batch variations that might occur when a new batch is used.

2. The major difficulty in the separation of most proteins of clinical interest, using FSCE, is the tendency for the proteins to be adsorbed onto the fused silica capillary wall, which causes broadening and tailing of peaks, resulting in marked loss in resolution and sensitivity. In the protocol described in this method, it was found that the use of a hydrophilic coated capillary, together with the addition of methyl cellulose to the electrophoresis buffer, has virtually eliminated the problem of protein adsorbtion.

3. The volume of CSF injected affects both the sensitivity and the separation of the constituents. In the authors' experience, a good overall separation of all the constituents in CSF is achieved with a sample injection time of 4 s. This is certainly the case for CSF proteins in fluids containing a total protein concentration of 200–500 mg/L. For samples that contain higher concentrations of total protein, better separations may be obtained by injecting less sample, i.e., 2 s for concentrations between 500 and 600 mg/L. Conversely, in the case of CSF obtained from the ventricles of the brain, in which the total protein concentration is much lower than in CSF taken from the lumbar region, it may be necessary to inject a larger volume of sample, i.e., 5–6 s.

4. For accurate quantitation of proteins and peptides in CSF, a detector wavelength of 214 nm should be used. This is the maximum absorption wavelength for peptide bonds; however, the sensitivity of detection is lower than at 200 nm.

5. Setting the detector wavelength at 200 nm provides the most sensitive detection. At this wavelength, all constituents contained in CSF are detected, regardless of whether they are proteins or other substances, such as organic acids.

6. At the start of a separation of CSF analytes, the baseline can vary a little, which may affect quantitation. However, after 5 min the baseline has stabilized, and it was therefore found that, by starting the integration at exactly 5 min after the start of the run (as indicated on the instrument display), results were more reliable and reproducible. This may not be necessary with more sophisticated integrators.

References

1. Lowenthal, A. *(1964) Agar Gel Electrophoresis in Neurology.* Elsevier, Amsterdam.
2. Cumings, J. N., Shortman, R. C., and Tooley, M. (1970) Polyacrylamide disc electrophoresis of cerebrospinal fluid and cerebral cyst fluids. *Clin. Chim. Acta.* **27,** 29–34.
3. Wiederkehr, F., Ogilvie, A., and Vonderschmitt, D. (1987) Cerebrospinal fluid proteins studied by two-dimensional gel electrophoresis and immunoblotting technique. *J. Neurochem.* **49,** 363–372.
4. Cowdrey, G., Gould, B., Rees, J., and Firth, G. (1990) Separation and detection of alkaline oligoclonal IgG bands in cerebrospinal fluid using immobilised pH gradients. *Electrophoresis* **11,** 813–818.
5. Cowdrey, G., Firth, M., and Firth, G. (1995) Separation of cerebrospinal fluid proteins using capillary electrophoresis: a potential method for the diagnosis of neurological disorders. *Electrophoresis* **16,** 1922–1926.

5

Immunosubtraction as a Means of Typing Monoclonal and Other Proteins in Serum and Urine

Stephen M. Palfrey

1. Introduction

Serum and urine protein electrophoresis are used primarily to screen for the presence of monoclonal proteins found in conditions such as myeloma, non-Hodgkin's lymphoma, macroglobulinemia, and so on. Having demonstrated the presence of an abnormal band, further testing is required to identify both the immunoglobulin heavy- and light-chain types (e.g., IgG κ). With conventional agarose gel or cellulose acetate electrophoresis, this secondary testing is either by immunofixation *(1–3)* or immunoelectrophoresis *(4)*. In both methods, serum or urine is electrophoresed, and antibodies to each of the immunoglobulin classes is reacted with the abnormal protein. Insoluble protein–antibody complexes are formed, which can be visualized by staining with dyes such as Coomassie blue. These methods are sensitive, but can be time-consuming and labor-intensive.

Free-solution capillary electrophoresis of serum *(5,6)* and urine proteins *(7)* can also be used to detect monoclonal bands. The direct addition of antibodies to the sample swamps the electophoretic pattern, because antibodies are themselves γ-globulins. It is first necessary to immobilize the antibodies, and then to add patient sample. After an incubation period, the sample is removed. If a monoclonal protein has reacted with the antibody, then the band will disappear from the sample on electrophoresis. This process has been termed immuno-subtraction *(8)*.

The coupling of peptides and proteins to polysaccharides *(9)* and cyanogen bromide activated Sepharose *(10)* are used in affinity chromatography. This chapter describes the preparation and use of antibodies immobilized on Sepharose, to identify monoclonal proteins in serum and urine. The

From: *Methods in Molecular Medicine, Vol 27: Clinical Applications of Capillary Electrophoresis*
Edited by: S. M. Palfrey © Humana Press Inc., Totowa, NJ

technique can also be used to detect other proteins, such as C-reactive protein (CRP), transferrin, and so on.

2. Materials
2.1. Reagents

1. CNBr-activated Sepharose 4B (Pharmacia Biotech, St. Albans,UK).
2. Coupling buffer: 4.2 g sodium bicarbonate, 2.9 g sodium chloride, dissolve and make to 100 mL with high-performance liquid chromatography (HPLC)-grade water. Adjust the solution to pH 8.8.
3. Blocking reagent: 1.5 g glycine dissolved in 100 mL HPLC water, and adjusted to pH 8.0.
4. Wash buffer A: 13.6 g sodium acetate and 29 g sodium chloride dissolved in HPLC water; make up to 1 L. Adjust the solution to pH 4.0.
5. Wash buffer B: 8.4 g sodium bicarbonate and 29 g sodium chloride dissolved in HPLC-grade water; and make up to 1 L. Adjust the pH to 8.4.
6. 1 mM HCl; Dilute 0.2 mL concentrated (10 M) HCl to 2 L with HPLC water, store at 4°C.
7. Saline: 4.5 g sodium chloride, 0.50 g sodium azide; dissolve in 500 mL HPLC grade water.
8. Immunofixation-grade antiserum for each protein of interest (*see* **Note 1**).

2.2. Equipment

1. Vacuum filtration system capable of filtering up to 1 L.
2. 0.45-µ cellulose acetate filter pads.

3. Method
3.1. Preparation of Immobilized Antisera

1. For each antibody, weigh out 0.5 g CNBr-activated Sepharose (*see* **Note 2**), e.g., if you are preparing IgG, IgA, IgM, κ, and λ antibodies, weigh out 2.5 g Sepharose.
2. Transfer the Sepharose to a 1-L conical flask, and add 500 mL cold (4°C) 1 mM HCl. Mix by gently swirling for 5 min to swell the Sepharose.
3. Vacuum filter using a .45-µ filter disk, continue to apply a vacuum, after the last of the HCl has been filtered, until the gel cake starts to crack.
4. Divide the gel between an appropriate number of screw-capped 10-mL glass tubes.
5. Dilute each antibody with an equal volume of coupling buffer, mix, and add to the gel. Use 1 vol of antibody, 1 vol of coupling buffer, and 1 vol of swollen gel. Cap the tubes, and mix for 2 h on an orbital mixer (*see* **Note 3**).
6. Centrifuge each tube at 500g for 1 min, and remove as much of the liquid phase as possible.
7. Fill the tube with blocking reagent, and return to the orbital mixer for 2 h (*see* **Note 4**). Centrifuge for 1 min at 500g, and remove as much of the liquid phase as possible.

8. Fill the tube with wash reagent A, mix by inversion, and centrifuge for 1 min at 500*g*. Remove as much of the liquid phase as possible.
9. Fill the tube with wash reagent B, mix by inversion, and centrifuge for 1 min at 500*g*. Remove as much of the liquid phase as possible.
10. Repeat steps 7–9 a further 3× (*see* **Note 5**).
11. Finally, wash the gel with saline, and store it under an equal volume of saline at 4°C.

3.2. Identification of Protein Bands by Immunosubtraction

Use a method for the electrophoresis of serum or urine proteins as described in Chapters 2 and 3. Serum samples are prediluted 50× with electrophoresis buffer prior to analysis. In the following method, this is referred to as the usual dilution.

1. Make the usual dilution of serum in electrophoresis buffer; do not dilute urines. If the actual concentration of the protein of interest is then greater than 0.4 g/L (*see* **Note 6**) make a further dilution in buffer, to bring it below this level.
2. Suspend the appropriate immobilized antibodies by inversion, and pipet 100 µL into a conical-bottomed plastic tube. Centrifuge each tube for 1 min at 500*g*, and remove all the liquid (*see* **Note 7**).
3. To each tube, add 75 µL of diluted serum or neat urine, mix gently, and cap the tubes. Leave for 1 h. Frequently resuspend the gel as it settles out (*see* **Notes 8** and **9**).
4. Centrifuge the tubes for 1 min at 500*g*, and transfer the supernatant to a CE sample vial.
5. Perform electrophoresis (without further sample dilution) in the usual way for each antibody-treated sample, and also include a reference sample that has not been treated with antibody.

3.3. Interpretation

1. Visual inspection of the electrophoresis pattern is usually sufficient to decide on the type of the monoclonal protein. If the protein is of the same type as the antibody, then the band will disappear (*see* **Note 10**).
2. **Figure 1** shows the electrophoresis of a serum sample from a patient with an IgG λ monoclone. It is easy to see that the large monoclonal band at 3.8 min in (A) has disappeared in the samples reacted with IgG (C) and λ (D) antibodies, but is still present in the sample reacted with IgA antibody (B). It was also still present after reaction with IgM and κ antibodies, but these are not shown.
3. **Figure 2** shows a urine positive for Bence Jones protein. The solid line is the sample reacted with κ light chain antibody, and the dotted line is the same sample reacted with λ light-chain antibody. The two electropherograms are slightly off-set, as can be seen from the urea peak at about 5 min. This would be typed as free λ light chains (*see* **Note 11**).
4. **Figure 3** is the superimposition of two electrophoreses. The solid line is the untreated sample, and the dotted line is a sample that has been reacted with immobilized C-reactive protein (CRP) antibody. The CRP peak (150 mg/L) is at 4.27 min, the antibody has completely removed this peak.

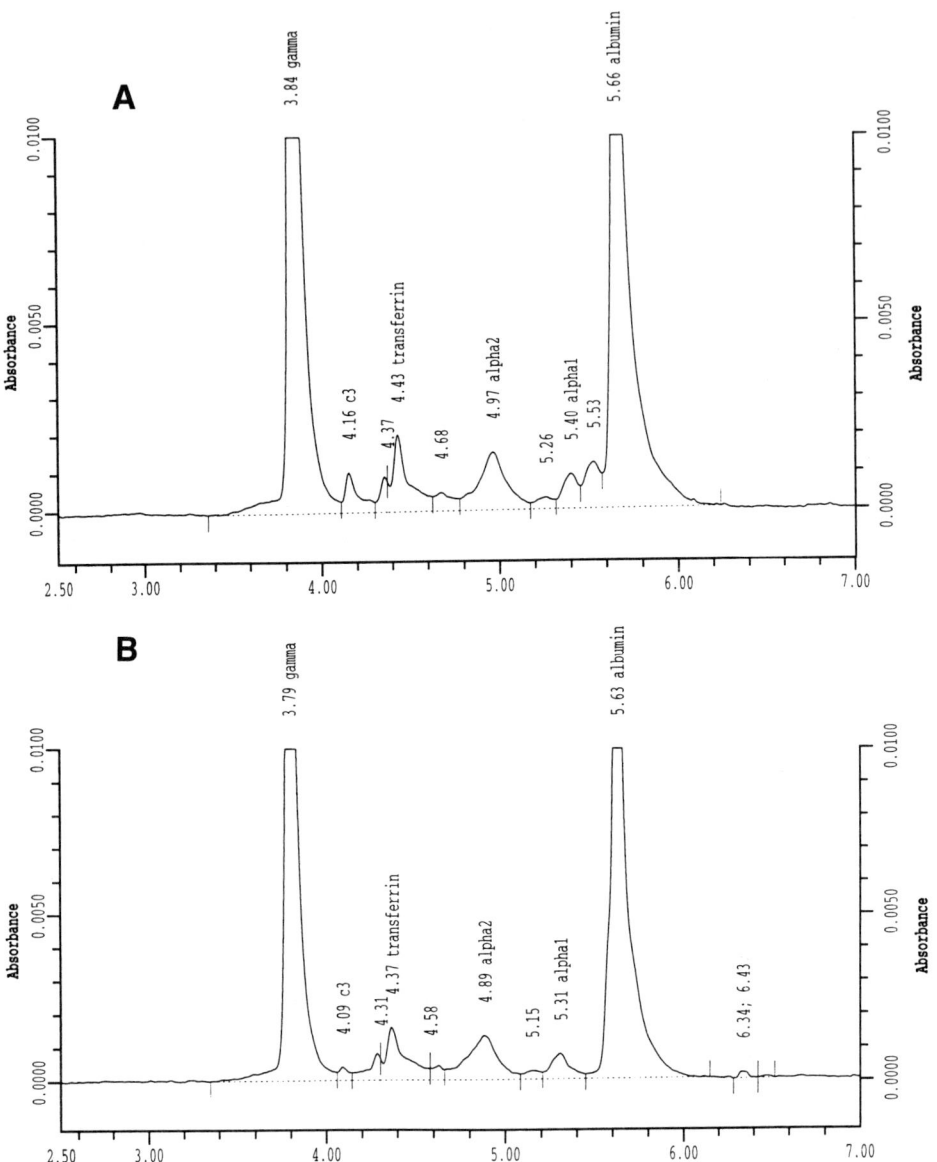

Fig. 1. (**A**) Electrophoresis of serum sample from a patient with an IgG 1 parapro-tein. (**B**) same sample after reaction with immobilized IgA antibody.

4. Notes

1. Other grades of anitisera can be used, but these often have many other proteins present that will also couple to the gel, resulting in a low-titer end product.

42

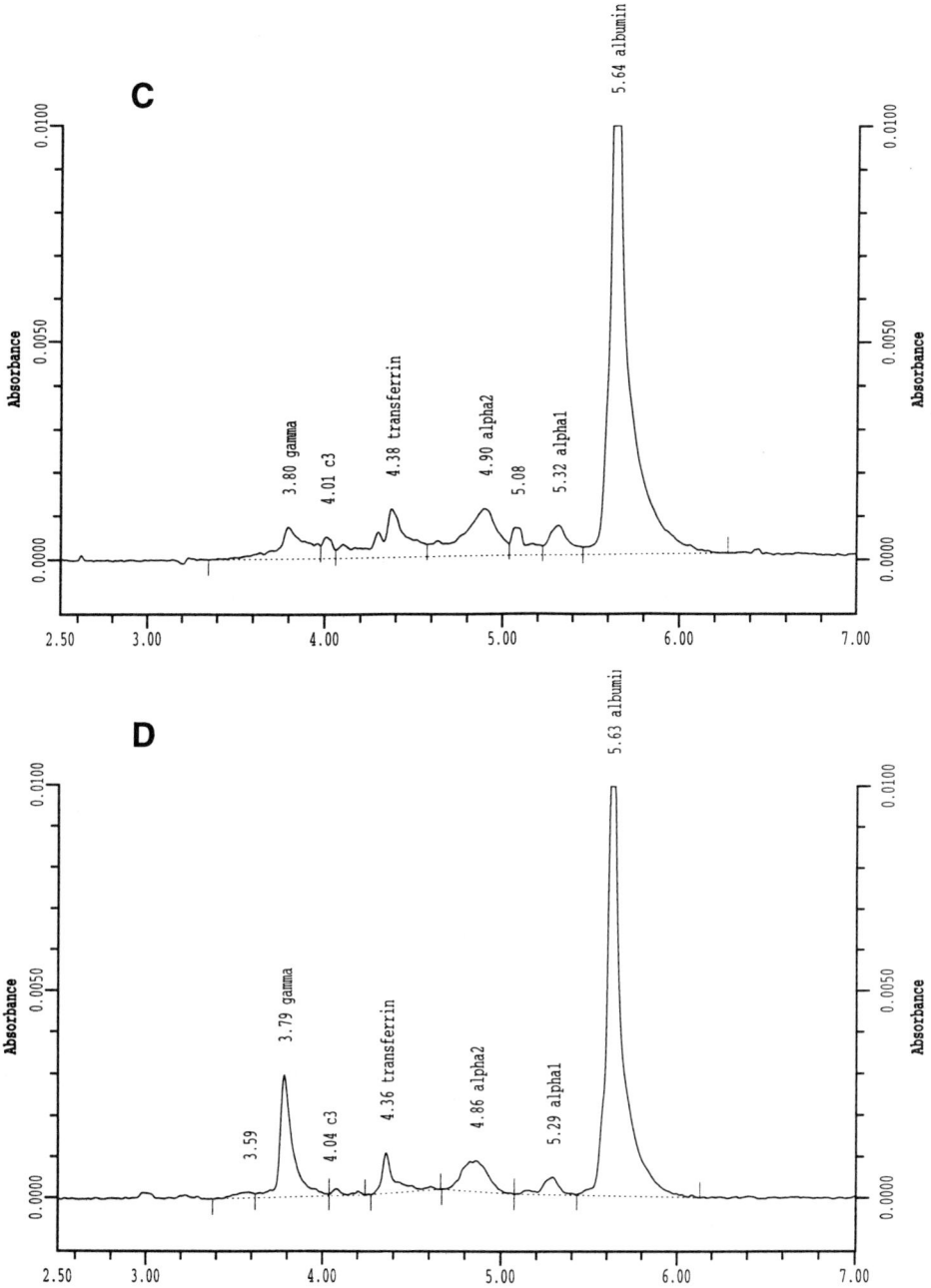

(**C**) after reaction with IgG antibody; (**D**) after reaction with Lambda light chain antibody.

2. 0.5g gel will yield approx 1.5 mL immobilized antibody.
3. Do not use a magnetic stirrer.

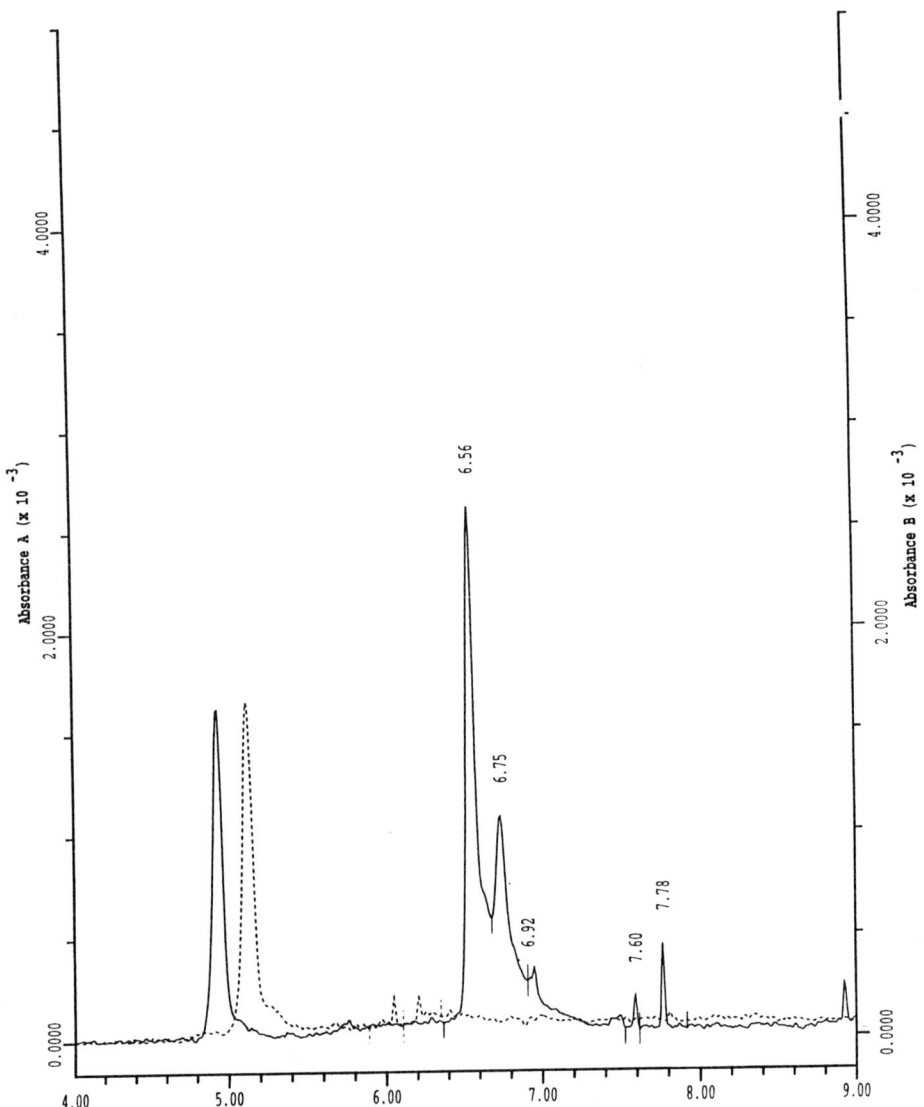

Fig. 2. Electrophoresis of a urine sample from a patient containing free lambda light chains, reacted with kappa light chain antibody (solid line), and after reaction with immobilized lambda light-chain antibody (dotted line).

4. The blocking reagent is used to react with any remaining activated sites on the gel. This prevents nonspecific binding of proteins in the sample to the gel.
5. This is an essential step: It ensures that no free ligand remains ionically bound to the immobilized ligand. The washing cycle of low and high pH is essential. Pro-

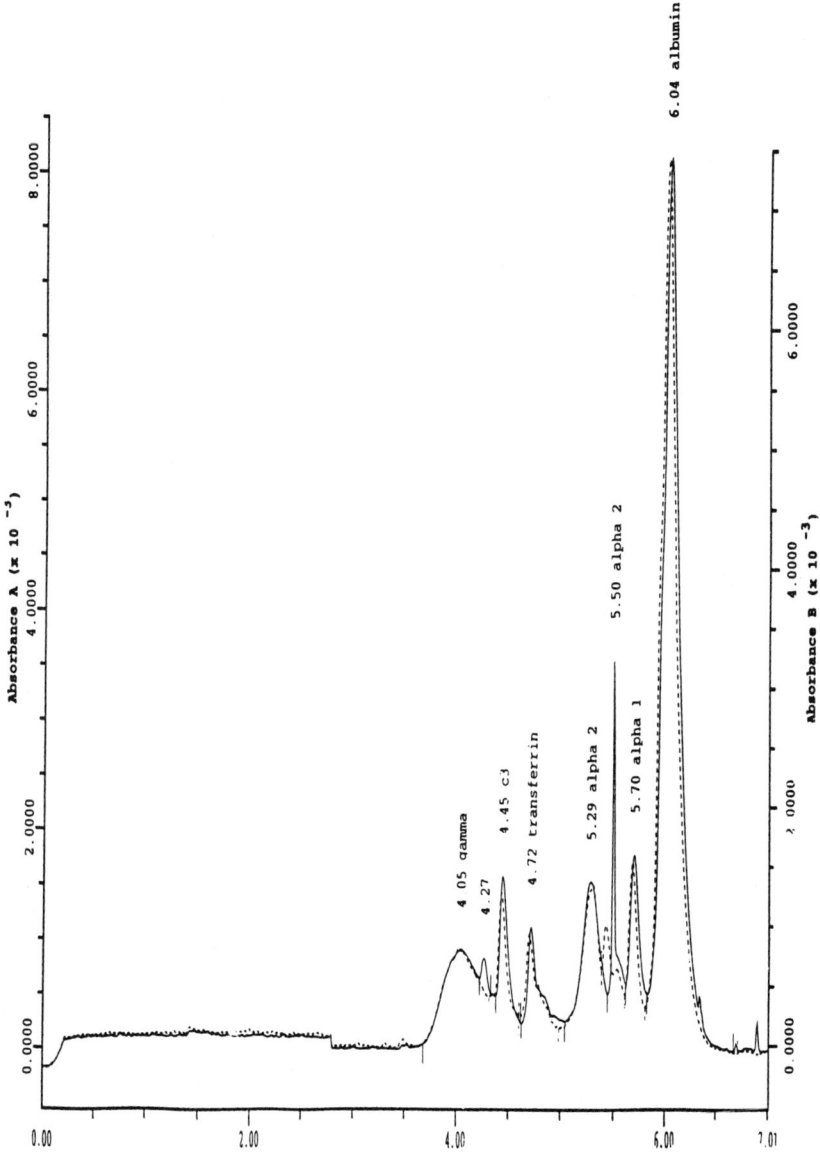

Fig. 3. Serum sample from a patient with a CRP of 150 mg/L (solid line), and after reaction with immobilized CRP antibody (dotted line). The CRP peak is at 4.27 min.

tein desorption occurs only when the pH is changed, pH changes do not cause loss of covalently bound protein.

6. This is an average figure, and may be greater or smaller, depending on the titer of the antibodies used.

7. The gel is very difficult to pipet, by suspending it in an equal volume of saline, twice the required volume can be pipeted easily. The saline can be removed after centrifugation.
8. The tubes can be placed on a flat bed roller mixer. This keeps the gel in suspension during the incubation, and improves the binding of antigen to the immobilized antibody.
9. Some proteins seem to react slowly with the immobilized antibody, particularly some IgM paraproteins, presumably because of steric effects at the binding site. Either use more solid-phase or prolong the incubation. The author has successfully used overnight incubations.
10. In practice, bands do not always disappear, but are usually markedly reduced in size (*see* **Note 9**). Occasionally, it is helpful, if the operating software allows, to superimpose electropherograms from different antibodies, to compare peak sizes.
11. To be certain that it is free light chains, you must either react the sample with a mix of IgG, A, and M antisera, and demonstrate no subtraction, or use immobilized antisera to free light chains.

References

1. Ritchie, R. F. and Smith, R. (1976) Immunofixation III. Application to the study of monoclonal proteins. *Clin. Chem.* **22,** 1982–1985.
2. Aguzzi, F. and Rezzani, A. (1986)An advantageous but neglected technique for immunofixation of monoclonal components on cellulose acetate membranes. *Giorno. Ital. Chim. Clin.* **1986,** 293–299.
3. Whicher, J. T., Hawkins, L., and Higginson, J. (1980) Clinical applications of immunofixation: a more sensitive technique for the detection of Bence Jones protein. *J. Clin. Pathol.* **33,** 779–780.
4. Grabar, P., and Williams, C. A., (1953) Methode permettant l'etude conjugee des proprietes electrophoretiques et immunochimiques d'un melange de proteines. Application au serum sanguin. *Biochim. Biophys. Acta* **10,** 193–194.
5. Kim, J. W., Park, J. H., Park, J. W., Doh, H. J. Heo, G. S., and Lee, K.-J. (1993) Quantitative analysis of serum proteins separated by capillary electrophoresis. *Clin. Chem.* **39,** 689–692.
6. Jenkins, M. A. and Guerin, M. D. (1995) Quantification of serum proteins using capillary electrophoresis. *Ann. Clin. Biochem.* **32,** 493–497.
7. Jenkins, M. A. (1997) Clinical application of capillary electrophoresis to unconcentrated human urine protein. *Electrophoresis* **18,** 1842–1846.
8. Khatter, N., Ashby, D., Alter, S., Mindeman, L., Kim, H., Wang, H., Liu, J., and Blessum, C. (1993) Identification of monoclonal components by immunosubtraction and capillary electrophoresis. *Clin. Chem.* **39,** 1136.
9. Axen, R., Porath, J., and Ernback, S. (1967) Chemical coupling of peptides and proteins to polysaccharides by means of cyanogen halides. *Nature* **214,** 1302–1304.
10. Pharmacia LKB Biotechnology. *Affinity Chromatography: Principles and Methods.*

6

Analysis and Classification of Serum Cryoglobulins

Zak K. Shihabi

1. Introduction

Cryoglobulins (CG) are immunoglobulins that reversibly precipitate from serum in cold temperatures. They are classified into three types, based on the monoclonality of the γ-globulins present (1): Type I contains only monoclonal bands; type II contains mixed polyclonal immunoglobulins with a monoclonal component; and type III contains mixed polyclonal immunoglobulins only.

CG can precipitate in the extremities, in the skin causing skin lesions (purpura); and in the kidney causing nephropathy (1,2). Type I produces large precipitates, and is associated with lymphoproliferative disorders, e.g., myeloma, and especially Waldenstrom's macroglobulinemia (1), producing symptoms of vasculitis or distal gangrene/necrosis. On the other hand, the mixed cryoglobulins (types II and III) produce small precipitates, and are associated with diseases stimulating the immune system (autoimmune, infections, and chronic inflammations) such as in chronic hepatitis (3,4), rheumatoid arthritis (5), and systemic lupus erythematosus (6,7), principally with symptoms of purpura, arthralgia, and Raynaud's phenomena. Type II is often associated with Sjögren's syndrome and chronic hepatitis C, with the cryoprecipitate of the infected patients being rich in hepatitis C virus (3,4). Type III is associated more with systemic lupus erythematosus and essential cryoglobulinemia (1).

Several methods are used for detection of the cryoglobulins. Capillary zone electrophoresis (CZE) can detect monoclonality of proteins, utilizing small amounts of proteins, and at the same time offers several new aspects to their analysis (8): good sensitivity at 214 nm, permitting a 10-fold decrease in sample size; good linearity, which allows direct quantification without further chemical reactions; and avoidance of extensive washing by using mathematical correction factors to compensate for any co-precipitation of serum proteins. In addition, the

From: *Methods in Molecular Medicine, Vol 27: Clinical Applications of Capillary Electrophoresis*
Edited by: S. M. Palfrey © Humana Press Inc., Totowa, NJ

speed, simplicity, and low operating cost render CZE an attractive method for analysis of CG in routine clinical laboratories. Here is a description of how CZE can be used to analyze and classify CG *(8)*.

2. Materials

1. Instrument: The author uses two instruments, a Model 2000 CE (Beckman, Fullerton, CA) and a model 4000 Quanta (Waters, Milford, MA). The electropherograms from the two instruments are similar.
2. Capillary: Use a 30 cm × 50 μm (id) untreated (uncoated) fused silica capillary. A new capillary is treated as described in the chapter on acetonitrile stacking (*see* Chapter 17, **Note 7**).
3. Electrophoresis buffer, pH 8.8: 7 g boric acid, 7 g anhydrous sodium carbonate, and 5 g polyethylene glycol 8000 in 1000 mL water.
4. Borate buffer 1: 40 m*M*, pH 9.0.
5. Borate buffer 2: 40 m*M*, pH 8.8.

3. Methods

3.1. Cryoglobulin Precipitation

1. Keep the collected blood warm at 37°C during transportation and centrifugation.
2. Remove the serum, and store at 1–4°C for a period of 5 d (*see* **Note 1**).
3. At the end of this period, mix the cooled serum well, and remove two aliquots for analysis.
4. Centrifuge 500 μL of one aliquot, while cold, at 15,000*g* for 10 s in a Beckman microcentrifuge.
5. Decant the supernatant carefully, and wash the precipitate (often difficult to see) with ice-cold distilled water, and centrifuge again.
6. Dissolve the precipitate in 100 μL warm (37°C) borate buffer 1, pH 9.0. Occasionally, a larger volume of buffer is needed to dissolve the precipitate (*see* **Notes 2** and **3**).
7. Transfer the sample to the CE system.
8. Warm the second aliquot, about 200 μL, at 37°C for 10 min, to dissolve any precipitate, and mix.
9. Dilute 10 μL with 500 μL of borate buffer 2, pH 8.8, and transfer to the CE system.

3.2. Electrophoresis

1. Set the voltage to 230 V/cm, and the detector to 214 nm.
2. Load the sample hydrodynamically (0.5 psi) for 10 s onto the capillary.
3. Electrophorese for 10 min.
4. After each run, rinse the capillary at high pressure (20 psi) for 2 min with 0.2 *M* NaOH, followed by electrophoresis buffer for 2 min (*see* **Note 4**).

3.2. Quantification

1. The area under the peak is used for quantification (*see* **Note 5**).
2. The amount of co-precipitated serum globulins is subtracted from the area under the γ-globulin peak of the CG, based on the amount of albumin present in that fraction, according to the formula:

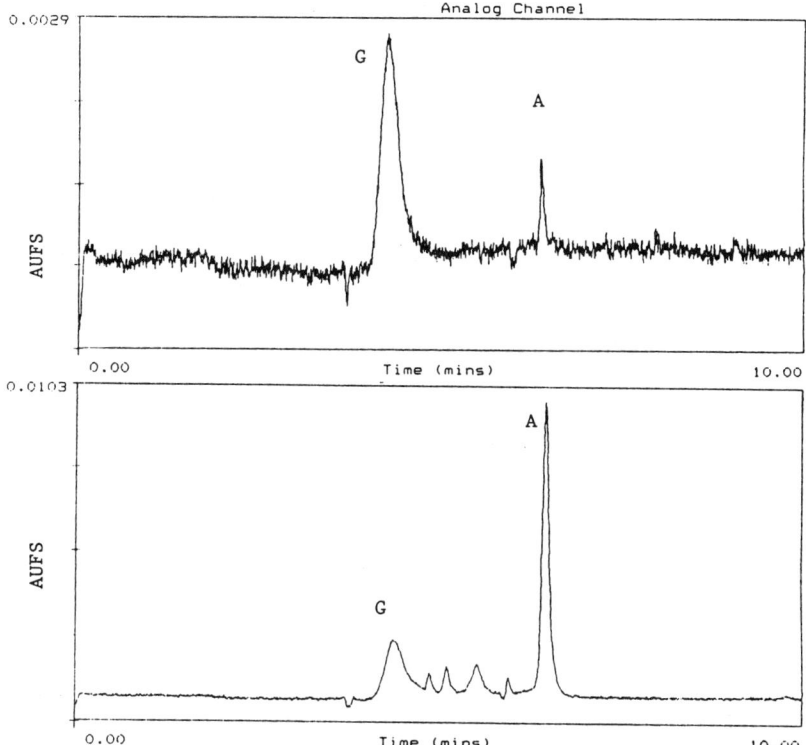

Fig. 1. (Top) Example of type III CG; (bottom) serum of the patient. Note small amount of albumin is present in CG; however, the γ region of the CG relative to albumin is much higher than that of the serum.

Total γ-globulins area under the CG peak
= ([area caused by CG] + [area caused by serum γ-globulins]) × (dilution)
= ([area caused by CG] + [albumin of CG × ratio of serum globulins/
   serum albumin]) × (dilution) (*see* **Note 6**).   (1)

3.3. Interpretation

1. The first step is to determine whether the sample is positive. A ratio of γ-globulins to albumin in the CG equalling that of the serum (warmed) is considered negative; a higher ratio is considered positive (indicates preferential cold precipitation of CG).
2. The second step is classification of the positive samples, based on detecting the presence of monoclonal band(s) in the γ-region, i.e., narrow or sharp peaks vs wide peaks, as illustrated in **Figs. 1** and **2** for types II and III (the common types). Type I (less common) shows a large sharp monoclonal peak in the γ region in the CG, with the same peak present also in the serum, but at a lower concentration relative to albumin (*8*).

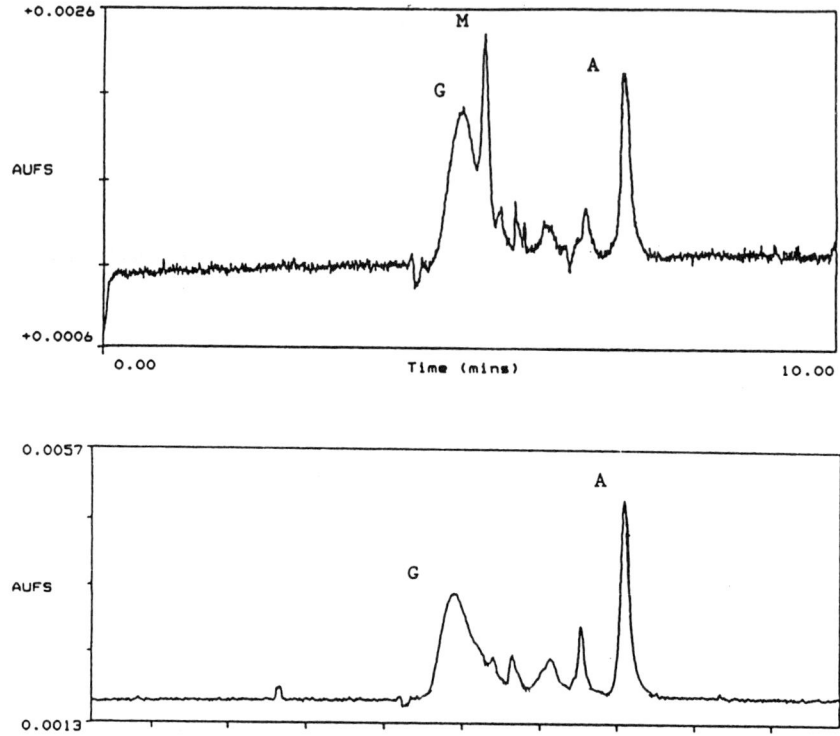

Fig. 2. (Top) Example of type II CG; (bottom) serum of the patient. Note the CG shows a monoclonal peak (M) which is absent from the serum. Note also the relative increase of the total γ region relative to albumin in the CG vs that for the serum.

3. It is important to run a serum protein electrophoresis of the patient for the proper interpretation and quantification of cryoglobulins.
4. This method also detects serum monoclonal spikes and oligoclonal bands.

4. Notes

1. Types I and II precipitate rapidly, within a few hours at 4°C, or on ice-cold water, forming visible precipitates. Type III requires about 5 d, and the precipitate is not visible.
2. Large precipitates may need to be solubilized in 500 μL, instead of 100 μL of the buffer.
3. Cryoglobulins are different from one individual to another, and thus they exhibit slightly different characteristics as far as solubility, precipitation, migration, and so on. Before introducing into the instrument, it is important to inspect the sample visually for complete solubility. Some samples do not dissolve easily and may require the addition of 5 μL of 1 *M* NaOH to the buffer, although this treatment

slightly denatures proteins *(9)*. Some cryoglobulins, especially IgM, form turbidity, causing falsely elevated values. These can be solubilized also by the sodium hydroxide.

4. Some very rare samples ruin the capillary, presumably because the proteins adher tightly to the capillary surface. In this case, the old capillary is replaced with a new one.

5. If the precipitate is heavy, it can be quantified as the difference between the total proteins in the warmed tube (no precipitate) and the total proteins in the supernatant of the cold tube.

6. In other words, albumin is used as a correction factor in the calculation of CG to eliminate contamination and co-precipitation from serum. This approach diminishes the need for extensive sample washing.

References

1. Brouet, J.-C., Clauvel, J.-P., Danon, F., Klein, M., and Seligman, M. (1974) Biologic and clinical significance of cryoglobulins: a report of 86 cases. *Am. J. Med.* **57,** 775–788.
2. Dolcher, M. P., Marchini, B., Sabbatini, A., Longombardo, G., Ferri, C., Riente, L., Bombardi, S., and Miglioarini, P. (1994) Autoantibodies from mixed cryoglobulinemia patients bind glomerular antigens. *Clin. Exp. Immun.* **96,** 317–322.
3. Agnello, V., Chung, R. T., and Kaplan, L. M. (1992) Role for hepatitis C virus infection in type II cryoglobulinemia. *N. Engl. J. Med.* **327,** 1490–1495.
4. Schifferli, J. A., French, L. E., and Tissot, J.-D. (1995) Hepatitis C virus infection, cryoglobulinemia, and glomerulonephritis. *Adv. Nephrol.* **24,** 107–109.
5. Tsuchiya, N., Malone, C., Hutt-Fletcher, L. M., and Williams, R. C. (1991) Rheumatoid factors react with Fab fragments of monoclonal antibodies to herpes simplex virus types 1 and 2 Fc gamma-binding proteins. *Arth. Rheumat.* **34,** 846–855.
6. Gripenberg, M., Teppo, A. M., Kurki, P., Gripenberg, G., and Helve, T. (1988) Autoantibody activity of cryoglobulins and sera in systemic lupus erythematosus. Association of IgM class rheumatoid factors with Raynaud's syndrome. *Scand. J. Rheumat.* **17,** 249–254.
7. Levo, Y. (1982) Cryoproteins, cryoimmunoglobulins, and cryofibrinogen. *Adv. Microcirc.* **10,** 73–94.
8. Shihabi, Z. K. (1996) Analysis and general classification of serum cryoglobulins by capillary zone electrophoresis. *Electrophoresis* **17,** 1607–1612.
9. Freidberg, M. A. and Shihabi, Z. K. (1997) Urine proteins analysis by capillary electrophoresis. *Electrophoresis* **18,** 1836–1841.

7

Myoglobin Analysis

Zak K. Shihabi

1. Introduction

Myoglobin represents the stores of oxygen in muscle tissues. Because of its relatively small mol wt, myoglobin is often used in electrophoretic techniques as a mol wt marker, and also as a test for separation efficiency in capillary electrophoresis (CE). The separation of myoglobin can be accomplished based on mol wt (1), based on charge, or both together (as used here).

From the clinical point of view, the presence of myoglobin in the urine occurs frequently in hospital patients as a result of the breakdown of muscle tissues, because of several causes, such as trauma and burns (2). It is important to recognize this condition as soon as possible, for early therapy to prevent acute renal failure (3–5). Hemoglobin (Hb) is the chief pigmented protein encountered in urine that can interfere in myoglobin detection. The separation of these two proteins sounds very simple, but in practice it is very difficult (5). Since myoglobin tends to denature rapidly in urine in a few hours, rapid methods, in terms of a few minutes, are needed for the analysis.

Several techniques have been advocated for the detection of myoglobin, including agarose electrophoresis, salt precipitation, immunoassays, and spectrophotometry, as summarized elsewhere (5). None of these is very practical for routine clinical use, because of interference, sensitivity, or time-consuming steps (5). The method described here uses CE to separate myoglobin from hemoglobin, and to quantify it directly in urine (6).

2. Materials

1. Capillary: Use a 25 cm × 50 μm (id) untreated fused silica capillary.
2. Instrument: A Model 2000 (Beckman, Fullerton, CA) equipped with a 405-nm filter.

From: *Methods in Molecular Medicine, Vol 27: Clinical Applications of Capillary Electrophoresis*
Edited by: S. M. Palfrey © Humana Press Inc., Totowa, NJ

3. Electrophoresis buffer: 150 m*M* borate, pH 8.2, with 0.5% polyethylene glycol (PEG) 8000 (Fisher Scientific,) (*see* **Notes 1** and **2**).
4. Standard: Horse myoglobin 100 mg/L of water, stable for about a week refrigerated (*see* **Note 3**).
5. Internal standard: Washed red blood cells were hemolyzed in water to give a solution of Hb ~2500 mg/L (stable for about a week refrigerated).

3. Methods
3.1. Tissue Extraction

1. Homogenize human skeletal muscle tissue samples from autopsy (200 mg) in 1.0 mL water.
2. Centrifuge for 1 min at 15,000*g* in a Beckman microcentrifuge. The supernatant is used for the analysis.

3.2. Procedure

1. Before using for the first time, condition a new capillary by rinsing with 0.2 *M* NaOH for 20 min, then with water for 2 min, followed by separation buffer for 5 min.
2. The voltage is set to 6 kV and the detector to 405 nm (*see* **Note 4**).
3. Introduce the sample by low pressure injection (0.5 psi) for 6 s.
4. Wash the capillary between injections with phosphoric acid, 7 m*M* for 40 s, followed by the electrophoresis buffer for 60 s (*see* **Note 5**).
5. Inject urine samples directly into the capillary as soon as possible after collection (*see* **Note 6**).
6. If a peak is present, mix a 5-µL aliquot of the internal standard hemolysate with 25 µL of the urine, and inject again into the capillary. The Hb peak is used as an internal standard for the migration time (*see* **Note 7**).

3.3. Interpretation

1. **Figure 1** illustrates the separation of horse and human myoglobin, and hemoglobin added to a urine sample from a normal individual (*see* **Note 8**)
2. **Figure 2** illustrates a urine sample from a patient with myoglobinuria.
3. This method detects values >20 mg/L. This level is adequate for clinical detection of myoglobinuria (**ref.4**; *see* **Notes 9–11**).
4. The test is linear between 20 and 400 mg/L.

4. Notes

1. Prepare the buffer by dissolving the correct amount of boric acid and adjusting the pH slowly to the desired pH with 2.5 *M* NaOH, without going above that value. If the pH goes over the desired pH, discard the buffer to avoid changes in the ionic strength. Ionic strength greatly affects the migration time in CE.
2. Zwitterionic surfactants, such as 5 m*M* n-alkyl dimethylamino-propane sulfonate, added to the separation buffer are said to improve peak shape for myoglobin and lysozyme, decreasing the adsorption of the proteins to the capillary walls (*10*).

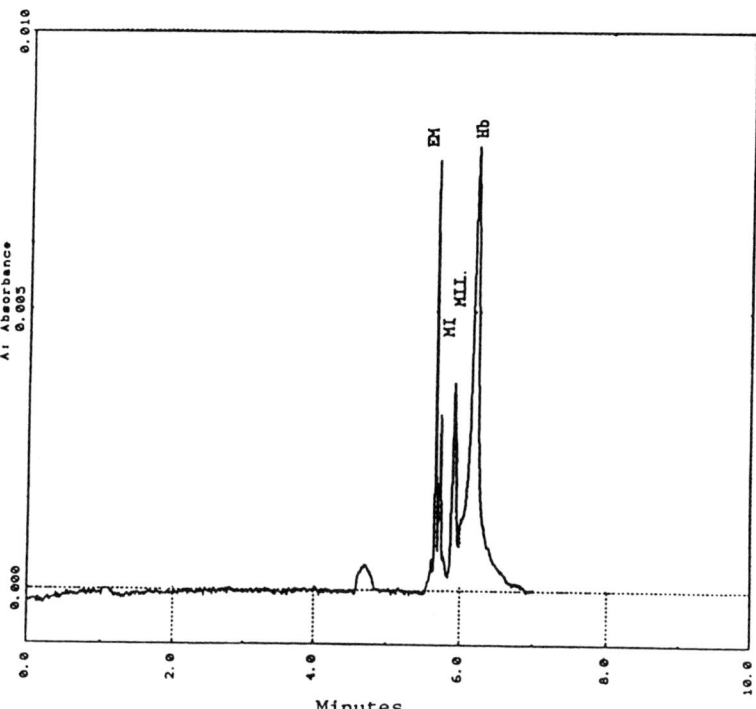

Fig. 1. Horse myoglobin (EM), human muscle myoglobin (MI), human muscle myoglobin (MII), and hemoglobin (Hb).

3. For practical reasons, horse myoglobin (Sigma, St. Louis, MO) is used as a standard for quantification. It migrates slightly ahead of human myoglobin.

4. The use of a 405-nm filter offers good sensitivity, and eliminates most of the interferences present in urine from other proteins, peptides, and small molecules. Hemoglobin is the peak most often detected in urine. Hemoglobin variants, such as S and C, do not interfere with the MI peak (the chief myoglobin peak in urine).

5. Over 100 samples can be injected onto the capillary without any major deterioration in the separation. The use of high ionic strength buffers, with acid washing between injections, contributes to good separation and long capillary life.

6. Myoglobin in urine is not stable, regardless of the storage conditions. Once the sample is denatured, its solubility decreases and it behaves differently in most separation media. Samples can be stored for ~4 h refrigerated.

7. The electrophoretic mobilities of myoglobin and hemoglobin change slightly, depending on the particular urine sample. They also differ from that of pure standards, because of the differences in salt, pH, viscosity, and so on, of the different urine samples. In order to avoid this problem, the urine sample is analyzed again after the addition of hemoglobin (as an internal standard), so that both proteins

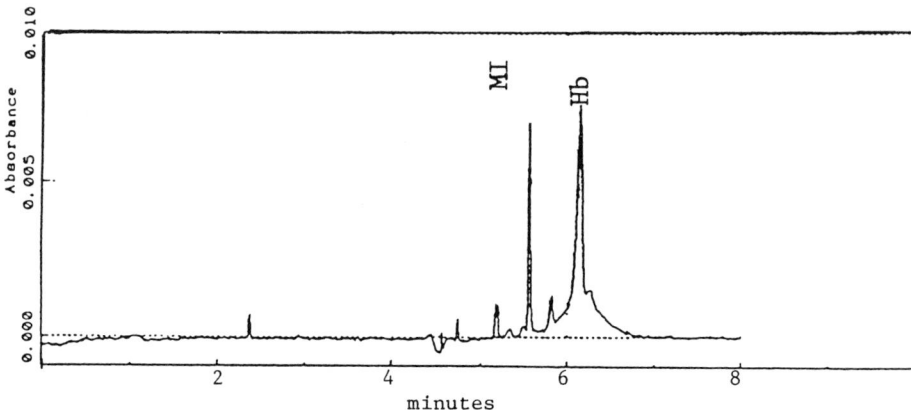

Fig. 2. Urine sample from a patient with myoglobinuria (340#mg/L). Hemoglobin (Hb) is added as an internal standard (MI, myoglobin).

will be affected in the same manner and have the same ratio of mobility *(5)*. Of course, it would be better to use the patient's hemoglobin as an internal standard, to avoid the presence of an occasional F, S, or C variant. These variants have slightly different mobilities, but do not interfere with the myoglobin peak. In addition, myoglobin has a distinct sharp peak different from the wider peak of hemoglobin.

8. Note that fresh extracted myoglobin from muscle tissues gives two peaks (MI and MII) by CE *(6)* and by HPLC *(7)*; however, principally one peak (MI) is present in the urine.

9. Myoglobin can be present in the urine and serum of a normal individual at concentrations <1 mg/L, which is below the detection of this method.

10. To increase the sensitivity of the method, concentration by membrane filtration, use of wider capillaries, or stacking on the capillary can be attempted.

11. Low levels of myoglobin, as well as other proteins, can be separated better in coated capillaries *(8,9)*; however, these capillaries differ greatly in their stability, and they require different conditions for separation.

References

1. Zhang, Y., Lee, H. K., and Lee, S. F. Y. (1996) Separation of myoglobin molecular mass marker using non-gel sieving capillary electrophoresis. *J. Chromatogr. A.* **744,** 249–258.
2. Kagen, L. J. (1973) *Myoglobin: Biochemical, Physiological and Clinical Aspects.* Columbia University Press, New York, p. 91–92.
3. Dubrow, A. and Flamenbaum, W. (1988) Acute renal failure with myoglobinuria and hemoglobinuria, in *Acute Renal Failure* (Brenner, B. M. and Lazarus, J. M., eds.), Churchill and Livingstone, New York, pp. 279–293.

4. Ward, M. M. (1988) Factors predictive of acute renal failure in rhabdomyolysis. *Arch. Int. Med.* **148,** 1553–1557

5. Hamilton, R. W., Hopkins, M. B., and Shihabi, Z. K. (1989) Myoglobinuria, hemoglobinuria and acute renal failure. *Clin. Chem.* **35,** 1713–1720.

6. Shihabi, Z. K. (1995) Myoglobinuria detection by capillary electrophoresis. *J. Chromatogr. B.* **669,** 53–58.

7. Powell, S. C., Frielander, E. R., and Shihabi, Z. K. (1984) Determination of myoglobin by high-performance liquid chromatography. *J. Chromatogr.* **317,** 87–92.

8. Nashabeh, W. and Rassi, Z. E. (1991) Capillary zone electrophoresis of proteins with hydrophilic fused-silica capillaries. *J. Chromatogr. A.* **559,** 367–383.

9. Schmalzing, D., Piggee, C. A., Foret, F., Carrilho, E., and Karger, B. L. (1993) Characterization and performance of neutral hydrophobic coating for the capillary electrophoretic separation of biopolymers. *J. Chromatogr. A.* **652,** 149–159.

10. Gong, B. Y. and Ho, J. W. (1997) Effect of zwitterionic surfactants on the separation of proteins by capillary electrophoresis. *Electrophoresis* **18,** 732–735.

8

Enzyme Analysis

Cathepsin D as an Example

Zak K. Shihabi

1. Introduction

In enzyme analysis, capillary electrophoresis (CE) offers the ease of product separation from the substrate, with the ability to use expensive reagents in microvolumes. In CE, enzymes can be measured either as mass (when they are present in high concentration) by direct light absorbency, or by catalytic activity. For example, the protease enzyme, savinase, which is used as an ingredient in washing powder, was determined directly by its absorbency at 200 nm *(1)*. For catalytic activity measurements, the substrate, the product, or both can be measured in CE without the need for coupling reactions. Because of the increased sensitivity, most CE methods measure enzymes by their catalytic activity on a substrate. To accomplish this by CE, several approaches have been used.

1.1. On-Line (Inside the Capillary) Incubation

Several workers have described different methods for the analysis of enzymes by performing the incubation step inside the capillary. For example, Bao and Regnier *(2)* and Alvia and Whitesides *(3)* have described enzymatic analysis of glucose-6-phosphate dehydrogenase by CE, in which the capillary is filled with the substrate, coenzyme, and the running buffer. The capillary is used as a microreactor. After injecting the enzyme and mixing it electrophoretically, the potential is turned to zero to allow for product accumulation. The potential is turned on again to separate the products from the substrate. Wu and Regnier *(4)* described a similar analysis for alkaline phosphatase and β-galactosidase, in which both of the enzymes are assayed simultaneously, using a

From: *Methods in Molecular Medicine, Vol 27: Clinical Applications of Capillary Electrophoresis*
Edited by: S. M. Palfrey © Humana Press Inc., Totowa, NJ

gel-filled capillary to decrease band broadening caused by product diffusion. On-line analysis has the advantages of increased sensitivity and of using microamounts of reagents, but the resulting chromatograms are difficult to quantify.

1.2. On-Line (Postcapillary) Reactor

Emmer and Roeraade (5) separated glucose-6-phosphatase and 6-phos-phogluconic dehydrogenase by CE, then added the substrate using a post-capillary reactor through a T-connection. The microvolumes used in CE pose several practical problems for postcolumn reaction. The reaction has to be fast enough for detection, and band broadening can occur easily.

1.3. Off-Line (Outside the Capillary) Incubation

Incubation outside the capillary is more practical and easier to perform than on-line analysis in CE. The enzyme and the substrate, in this technique, are mixed and incubated outside the capillary. The CE is used mostly to separate the reaction products. Both the products and the reactants can be monitored simultaneously. The incubation step can be short or long without any restriction. Since the instrument can measure 0.00001 absorbency unit, good sensitivity can be achieved. For example, Landers et al. (6) have measured the enzyme chloramphenicol acetyl transferase activity by CE. The enzyme and the substrate were incubated outside the capillary and the products, acetyl chloramphenicol and CoA, were separated by CE. Glutathione peroxidase was measured in cell-free preparations by incubating the enzyme and the substrate outside the capillary, followed by separation of the reduced and oxidized glutathione by CE (7). The method compared well to an HPLC and to a coupled enzyme assay (7). The author assayed the enzyme N-acetyl-glucosamindase in urine and tissues by incubating the enzyme with the substrate outside the capillary, and separating the reaction product by CE (8).

One especially practical aspect in this approach is stopping the reaction. One way is by bringing the reaction to a low temperature on ice before injection on the capillary. However, it was found, in most instances, that using acetonitrile is more suitable for CE. In this situation, the acetonitrile has several advantages: It stops the reaction, removes the protein, overcomes the matrix effect, and produces stacking (*see* Chapter 17).

Proteolytic enzymes with low activity are well suited for analysis by CE in this manner. Enzymes such as the peptidases are suitable for analysis by CE, because the peptide bond can be detected at 215 nm without staining or the need for using labeled substrates. Angiotension-converting enzyme (9) and carboxypeptidase Y (10) were assayed by CE. A tripeptidase from lactococ-cus lactis was measured in about 10 min with CE without derivatization (11). The enzyme was reacted outside the capillary with the substrate gly-gly-phe, and the products were separated using a citrate buffer for electrophoresis. The CE was more useful than the traditional colorimetric assay in the detection of other contaminating enzymes *(11)*. Furthermore, CE can give basic information on the structure of the peptide, such as ionization, charge-to-mass ratio, and so on.

The analysis of cathepsin D, a proteolytic lysosomal enzyme with an optimum pH of 3.5, is described here *(12)*. Cathepsin D is secreted from some tumor cells, and aids in metastasis. Tissue enzyme levels are a good predictor of tumor malignancy in general, and of breast carcinoma in particular *(13)*. Initially, the enzyme was assayed by its catalytic activity on several proteins, and, more recently, by immunoassay *(13)*. Both of these methods are time-consuming (requiring about a day to perform) and expensive. Here, this enzyme is measured by its catalytic activity outside the capillary. After incubating the tissue homogenates with the substrate hemoglobin, acetonitrile was added to stop the reaction and precipitate the hemoglobin. The tubes are centrifuged and the supernatant is injected into the capillary. A specific peptide, which is soluble in acetonitrile, is cleaved, separated by capillary zone electrophoresis (CZE), and detected at 214 nm in less than 5 min. This peptide is not produced by the action of pepsin or trypsin. The test compares well to a radioenzymatic immunoassay *(12)*. The method demonstrates the advantages of CZE for assay of proteolytic enzymes in general.

2. Materials

1. Cathepsin buffer: Formic acid 0.3 M, pH 3.4, containing 12.5 g/L NaCl and 0.5 mL/L of Triton X-100 (stable at room temperature).
2. Hemoglobin solution: Lyophilized human hemoglobin (*see* **Note 1**) 0.24 g/10 mL in water (Sigma, St. Louis, MO). Keep this solution on ice.
3. Substrate: Equal volumes of hemoglobin solution and cathepsin buffer are mixed and kept on ice.
4. Deproteinization reagent: 7 mg Iothalamic acid (Malinckrodt, St. Louis, MO), used as an internal standard, is dissolved in 100 mL acetonitrile (*see* **Note 2**). Stable at room temperature.
5. Electrophoresis buffer: 175 mM boric acid, adjusted to pH 8.4 with 4 M NH$_4$OH.
6. Capillary: Untreated fused silica capillary 42 cm × 50 µm id.
7. Instrument: A Model 2000 capillary electrophoresis instrument (Beckman, Fullerton, CA) is set at 360 V/cm, 24°C and 214 nm.

3. Methods
3.1. Preparation of Tissue Homogenate

The method described here has been published previously *(13)*.

1. Using a scalpel, slice about 0.3–0.5 g of frozen tissue into thin sections.
2. Mince and add 4 mL Tris buffer, pH 7.4.
3. Homogenize the tissue by a polytron to a thick liquid, while keeping on ice.
4. Spin the homogenates at 140,000 g for 30 min. Store the supernatant frozen at – 20°C prior to use.
5. Analyze the supernatant for protein content, using the commercial reagent from Bio-Rad. (Hercules, CA).
6. Dilute the homogenates 10-fold with water. Mix 100 μL of the diluted homogenates with 5 mL of the dye.
7. Incubate for 5 min at room temperature, and read the tubes at 595 nm, using bovine albumin as a standard.

3.2. Enzymatic Reaction

1. Add 150 μL substrate to a 1.5-mL centrifuge tube, and bring to 37°C (*see* **Note 3**).
2. Add 50 μL tissue homogenate, calibrator (*see* **Note 4**, or control to the warmed substrate, and incubate for 20 min at 37°C.
3. At the end of the incubation period (*see* **Note 5**), add 500 μL deproteinizing reagent. Mix the tube contents for 15 s, and centrifuge at 14,000 g for 30 s (*see* **Note 6**).

3.3. Electrophoresis

1. Condition a new capillary by rinsing with 0.2 *M* NaOH for 20 min, then with water for 2 min, followed by separation buffer for 5 min.
2. Introduce the sample by low pressure injection for 10 s (2.5% of the capillary). Samples should be analyzed within 1 h.
3. After each run, wash the capillary for 1 min with NaOH 0.2 M/L, followed by electrophoresis buffer for 1.0 min.
4. **Figure 1** illustrates the enzymatic activity of cathepsin D from breast tissue at 0 and 20 min. A specific peptide, soluble in acetonitrile, is cleaved from the hemoglobin molecule, and migrates at ~2 min (*see* **Notes 7** and **8**).

4. Notes

1. Bovine hemoglobin is less expensive, but also gives lower activity than human hemoglobin.
2. Acetonitrile precipitates the hemoglobin and many other large peptides.
3. Among many different brands of microtubes one was found that interfered in the reaction by causing an unusual precipitate.
4. To transfer peak heights into units of activity, use a sample with known activity (i.e., analyzed by a different method) as a standard.
5. The method as described can measure values as low as 10 pmol/mg protein. However, it can be modified for much greater sensitivity, e.g., by longer incubation, wider capillary diameter, or by increasing the sample size.
6. Because acetonitrile evaporates rapidly, the samples needs to be kept closed.

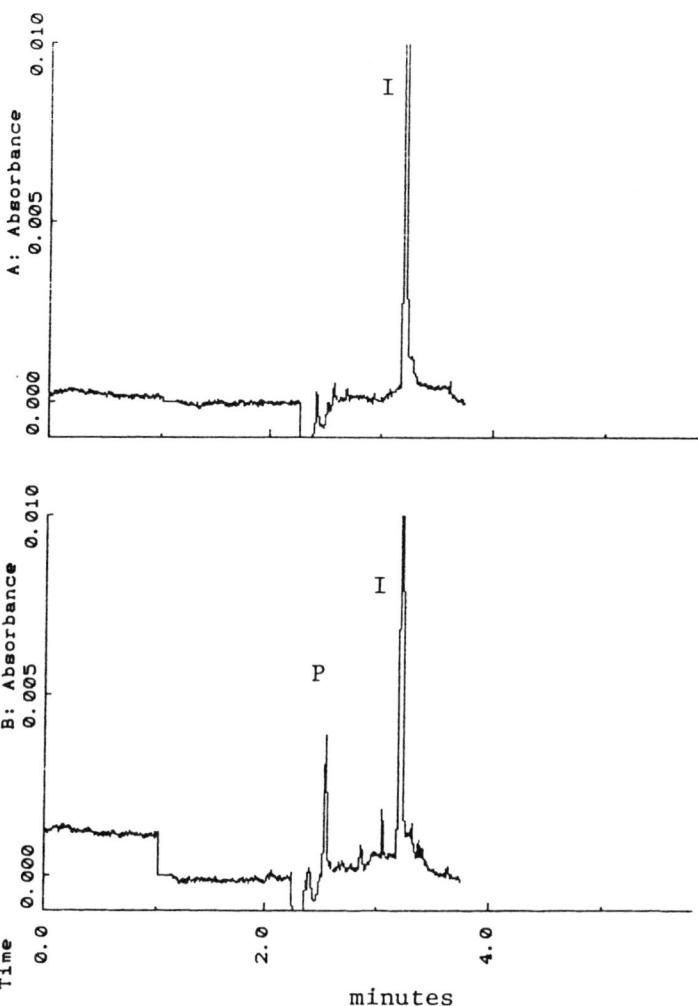

Fig. 1. Enzymatic activity of a sample of breast tumor homogenate (29 pmol/mg protein) at 0 (top) and 20 min after incubation (bottom). (P, split peptide; I, Iothalamic acid)

7. Interferences were not encountered. If interference is suspected, the sample can be analyzed at 0 time as a blank.
8. Initially, it is helpful to run the sample at 0 and 20 min after incubation, to be sure of the release of the peptide.

References

1. Vinther, A., Petersen, J., and Soeberg, H. (1992) Capillary electrophoretic determination of the protease savinase in cultivation broth. *J. Chromatogr.* **608,** 205–210.

2. Bao, J. and Regnier, F. E. (1992) Ultramicro enzyme assay in capillary electrophoretic system. *J. Chromatogr.* **608,** 217–224.

3. Avila, L. Z. and Whitesides, G. M. (1993) Catalytic enzyme activity during capillary electrophoresis: an enzymatic "microreactor" *J. Org. Chem.* **58,** 5508–5512.

4. Wu, D. and Regnier, F. E. (1992) Native protein separations and enzyme microassays by capillary zone and gel electrophoresis. *Anal. Chem.* **65,** 2029–2035.

5. Emmer, A. and Roeraade, J. (1994) Capillary electrophoresis combined with an on-line micro post-column reactor. *J. Chromatogr.* **662,** 375–381.

6. Landers, J. P., Schuchard, M. D., Subramaniam, M., Sismelich, T. P., and Spelsberg, T. C. (1992) High-performance capillary electrophoretic analysis of chloramphenicol acetyl transferase activity. *J. Chromatogr.* **603,** 247–257.

7. Pascual, P., Martinez-Lara, E., Barcena, J. A., Lopez-Berea, J., and Toribio, F. (1992) Direct assay of glutathione peroxidase activity using high-performance capillary electrophoresis. *J. Chromatogr.* **581,** 49–56.

8. Friedberg, M. and Shihabi, Z. K. (1997) Analysis of urinary N-acetyl-beta-glucosaminidase by capillary zone electrophoresis. *J. Chromatogr. B.* **695,** 187–191.

9. Shihabi, Z. K. (1992) Clinical application of capillary electrophoresis. *Ann. Clin. Lab. Sci.* **22,** 398–405.

10. Vinther, A., Adelhorst, K., and Kirk, O. (1993) Using capillary electrophoresis in the optimization of a carboxypeptidase Y catalyzed transpeptidation reaction. *Electrophoresis* **14,** 486–491.

11. Mulholland, F., Movahedi, S., Hague, G. R., and Kasumi, T. (1993) Monitoring tripeptidase activity using capillary electrophoresis Comparison with the ninhydrin assay. *J. Chromatogr.* **636,** 63–68.

12. Shihabi, Z. K. and Kute, T. (1996) Analysis of cathepsin D from breast tissues by capillary electrophoresis. *J. Chromatogr. B* **683,** 125–131.

13. Kute, T. E., Shao, Z. M., Sugg, N. K., Long, R. T., Russell, G. B., and Case, L. D. (1992) Cathepsin D as a prognostic indicator for node-negative breast cancer patients using both immunoassays and enzymatic assays. *Cancer Res.* **52,** 5198–5203.

9

Quantification of Human Cytomegalovirus by Competitive PCR and Capillary Electrophoresis

Zhongxin Yu, W. Douglas Scheer, and James M. Hempe

1. Introduction

Human cytomegalovirus (CMV) is a ubiquitous pathogen found in 40–100% of adults, and in about 1% of live births in the United States (1). It is the most common fetal and perinatal infectious organism; approx 10% of infected neonates are born with symptomatic congenital CMV disease, which is the most common cause of mental retardation and childhood deafness. CMV is a significant pathogen in immunocompromised individuals, including organ transplant recipients (2–4), and in acquired immune deficiency syndrome (AIDS) patients (5,6). Infection is characterized by latency, chronic infection, and reactivation, a progression similar to that observed in other members of the herpesvirus family. Because CMV infection is usually controlled by the host cellular immune system, primary infections can occur without obvious symptoms, and progress to latency may go unnoticed. Latent infection may persist throughout life, but primary or reactivated infection, coupled with impaired host immune response, can rapidly produce symptomatic CMV disease.

1.1. Laboratory Analysis

Traditionally, qualitative procedures, such as cell culture and serological assays, have been used for clinical assessment of CMV infection (7). However, because either viral load or a change in viral load may be directly related to the pathogenicity of CMV infection (8), quantitative assessment is needed for clinical evaluation of patients to confirm the onset of symptomatic CMV disease and/or to monitor the efficacy of antiviral therapy. The half tissue infectious dose ($TCID_{50}$) is a common quantitative assay in clinical use, despite the long turn-around time, relative insensitivity, and propensity for false positives. More

From: *Methods in Molecular Medicine, Vol 27: Clinical Applications of Capillary Electrophoresis*
Edited by: S. M. Palfrey © Humana Press Inc., Totowa, NJ

recent immunological methods, such as the shell vial assay for CMV immediate early antigen or lymphocyte antigen detection procedures, have also found widespread clinical application. Collectively, however, these assays lack specificity and sensitivity at low concentrations of CMV that may be clinically important.

DNA amplification by polymerase chain reaction (PCR) has induced dramatic advances in the molecular diagnostics of infectious diseases. Competitive PCR has been successfully applied to CMV quantification in plasma or serum, biopsy samples, peripheral blood leukocytes, urine, feces, and cerebrospinal fluid *(5–9)*. However, traditional methods for analysis of PCR products by slab-gel electrophoresis are complicated, labor-intensive, and difficult to quantify, so that they are not easily adapted for clinical use. This chapter describes a method that combines the specificity and sensitivity of viral DNA amplification by competitive PCR with sensitive automated detection of DNA amplicons by capillary electrophoresis (CE).

1.2. Method Overview

The described method takes two working days to measure CMV DNA in plasma. It can be adapted for urine, CSF, peripheral blood leukocytes, or other tissues (different DNA extraction procedures may be required). On d1, DNA in plasma from a negative control, two positive controls, and each patient, sample is extracted by silica gel affinity (~4 h). The extracted DNA is then amplified in five PCR reactions for each sample in the presence of 0, 10, 100, 1000, or 10,000 copies of internal standard (IS) DNA (~3 h). The amplification is competitive because the IS is plasmid DNA containing a 263-bp DNA insert with a nucleotide sequence that is identical to the amplification target, except for the presence of a *Bam*HI restriction site that was introduced into the sequence by site-directed mutagenesis. Enzymatic digestion of the IS amplicons with *Bam*HI, after PCR on d 2 (~4 h), yields a shorter fragment (245 bp) that can be separated from the CMV amplicon (263 bp) by capillary gel electrophoresis (CGE, 9 min assay, 45 min/sample).

CGE is performed using a 100 μm id × 27 cm long DB-1 capillary with electrophoresis buffer containing hydroxypropylmethylcellulose (HPMC) and SYBR Green I.(Molecular Probes, Eugene,OR) molecular dye, and amplicon detection by laser-induced fluorescence (LIF). Double-stranded DNA (dsDNA) in the PCR sample is electrokinetically injected into the capillary inlet, which is then immersed in electrophoresis buffer. The capillary outlet is simultaneously immersed in electrophoresis buffer at the cathode, and 8.1 kV is applied across the electrodes, generating an electric field of 300 V/cm. Current through the system increases rapidly to ~28 μAmps, and holds steady for the remainder of the separation. The IS and CMV amplicons pass the detector in less than 7 min. Peak area is measured in relative fluorescence units (RFU), and the number of copies of CMV DNA in a sample is determined by linear regression of the log of the ratio of the CMV peak area/corrected IS peak area against the log of the number of starting copies of IS DNA in the four samples. The point on the regression line at which the CMV peak area equals the corrected IS peak area (the log of the peak area ratio equals zero) is the number of copies of CMV DNA in the plasma sample.

2. Materials

Analytical grade reagents should be used unless otherwise noted. Suitable performance of all new lots of reagents and materials should be verified before reporting clinical results. At a minimum, the amount of CMV in the normal and abnormal controls should be within expected tolerance limits previously defined by replicate analysis.

2.1. Quantitative Competitive PCR

2.1.1. Plasma DNA Extraction

1. TLE buffer: 10 mM Tris-HCl, 0.1 mM/L EDTA disodium salt.
2. Silica suspension *(10)*: Suspend 60 g silica dioxide (SiO$_2$, 0.5–1 µm, Sigma, St. Louis, MO) in 500 mL deionized water in a graduated cylinder. Let stand until sedimentation is complete (~24 h), then aspirate the supernatant. Resuspend the silica in 500 mL deionized water a second time. After 5 h, aspirate the supernatant until 60 mL remains. Transfer to a 100-mL beaker, add 519 µL HCl (37% w/v), and mix thoroughly. Keep the silica particles in suspension by stirring, transfer 2-mL aliquots to 10-mL Pyrex tubes, cap loosely, and autoclave for 20 min at 121°C. After autoclaving, transfer the suspension to 0.5 mL microcentrifuge tubes, and store indefinitely at –20°C.
3. Lysis buffer: Add 60 g guanidine thiocyanate (GuSCN, Fluka Chemie AG, Buchs, Switzerland) to 50 mL 0.1 M Tris buffer, pH 6.4. Add 11 mL 0.2 mM EDTA, pH 8.0, and 1.4 mL Triton X-100 to the solution. If necessary, heat at 60°C to dissolve. Store up to 3 wk at room temperature in an opaque container.
4. Wash buffer: Add 60 g GuSCN to 50 mL 0.1 M Tris buffer, pH 6.4. If necessary, heat at 60°C to dissolve. Store up to 3 wk at room temperature in an opaque container.

2.1.2. Quantitative PCR

1. IS solutions: The IS used in this laboratory is plasmid DNA containing a 263-bp dsDNA insert that is a modified sequence of the amplification target. For competitive PCR, four IS solutions should be prepared in TLE buffer containing 10, 100, 1000, or 10,000 copies of IS/10 µL (*see* **Note 1**).
2. PCR reagent kit (Promega, Madison, MI): Contains Taq DNA polymerase (5 U/µL), 25 mM MgCl$_2$ and 10X PCR buffer, including 100 mM Tris-HCl, pH 9.0, at 25°C, 500 mM KCl, and 1% Triton X-100. Store at –20°C.

3. Nucleotide mixture (5X dNTP): Transfer 10 μL stock dATP, dGTP, dCTP, and dTTP (100 m*M*, New England BioLab, Beverly, MA) to a 1.5-mL microcentrifuge tube, and add 960 μL deionized water. Mix well, and store indefinitely at –20°C.

4. PCR primers: Sequences were based on cytomegalovirus strain AD169 *(11)*. Primers were prepared by LSUMC Core Laboratory (New Orleans, LA).
   a. Primer IE-1 (anneals to the coding region of the immediate early gene at positions 171431–171450, GenBank accession number Xl 7403): 5'-AGA CCT TCA TGC AGA TCT CC-3'.
   b. Primer IE-2 (anneals to the coding region of the immediate early gene at positions 171679–171698, GenBank accession number Xl 7403): 5'-GAC AAG GTG CTC ACG CAC AT-3'.

5. PCR master mix A: The required volume to prepare each day depends on the number of patient samples and controls to be analyzed. Multiply the following volumes by the number of samples to be analyzed (i.e., patient samples plus controls, include overage to account for transfer losses): 45 μL 10X Promega buffer II; 100 μL 25 m*M* MgCl$_2$ (*see* **Note 2**); 100 μL 5X dNTP; 50 μL primer IE-1 (10 μ*M*); 50 μL primer IE-2 (10 μ*M*); 5 μL deionized water.

6. PCR master mix B: The required volume to prepare each day depends on the number of patient samples and controls to be analyzed. Multiply the following volumes by the number of samples to be analyzed (*see* above): 40 μL type I H$_2$0; 5 μL 10X Promega buffer II; 5 μL Taq DNA polymerase.

2.2. Capillary Electrophoresis

1. DB-1 capillary: J&W Scientific, Folsom, CA. 100 μm id, 0.1-μm coating thickness (*see* **Note 3**).

2. TBE buffer: 89 m*M*/L Tris-HCl, 89 m*M*/L boric acid, 2 m*M*/L EDTA disodium salt, pH approx 8.6 (do not adjust).

3. Gel buffer (0.5% HPMC in TBE buffer): Add 250 mL TBE buffer to a 1 L glass bottle, and heat to 65 ± 5°C (~90 s in a microwave oven). Add 1.25 g HPMC (viscosity of 2% solution, 3500–5600 centipoises at 20°C, Sigma, St. Louis, MO), place the bottle on a stir plate, and mix for 10 min. Partially submerge the bottle in a container of ice water, and mix on a stir plate until the solution reaches 5–10°C (~30 min). Remove the bottle from the ice water (the solution should be clear and free of particles), and mix until the solution reaches room temperature (~25°C, 30 min). Vacuum-filter (0.45 μm, more than one filter may be required) and store up to 6 mo at 4°C.

4. Electrophoresis buffer (0.005% SYBR Green I in gel buffer): Store 2-μL aliquots of SYBR Green I stock solution (10,000X, proprietary concentration) at –20°C. Before each analytical run, prepare a working solution of SYBR Green 1(1:100 dye dilution) by adding 198 μL deionized water to the 2-μL aliquot. To make electrophoresis buffer, add 50 μL working solution to 10 mL gel buffer (1:20,000 dye dilution).

5. Membrane filters: 0.025-μm pore size, 25-mm diameter (Millipore, Bedford, MA cat. no. VSWP 025 00).

6. A dog nail clipper from a pet store is invaluable for converting 0.5-mL microcentrifuge tubes into topless sample vials.

3. Methods

The described method is based on previously published research *(10–12)*, as modified in this laboratory. Some degree of familiarity with PCR and CE instrumentation and operation is assumed. Procedural steps for CE are specifically described, as used with a P/ACE 5000 instrument with a laser induced fluorescence detector and version 8.1 of System Gold Software (Beckman, Fullerton, CA). Adaptation of these instructions to other platforms should be straightforward. Before reporting clinical results, each laboratory should validate their method by demonstrating suitable sensitivity, specificity, precision, linearity, and correlation with approved diagnostic methods.

3.1. Preparation of Unassayed Controls

Three plasma control samples should be evaluated with each analytical run to assure suitable PCR and CE performance. The use of frozen human plasma, rather than extracted DNA, is recommended so that quality control is verified for the entire analytical process. The negative control should be fresh plasma that has been shown to contain no CMV by both serological and PCR methods. Two positive controls spanning the pathological range of CMV infection should be prepared by adding cultured CMV to the plasma used for the negative control. The controls are stored as 200-µL aliquots in 15-mL polypropylene tubes at –70°C. This laboratory prepares controls containing approx 75 and 750 copies of CMV DNA/20 µL of plasma, and routinely obtains interassay CV less than 15%.

The authors also use a separate CE DNA control to verify CE instrument performance, independent of the rest of the analytical process. This control was prepared from the PCR product amplified from a positive plasma sample in which the IS and CMV DNA were empirically shown to be present in similar concentrations. The sample was digested with *Bam*HI, desalted, diluted 1000-fold in deionized water, and stored as 25-µL aliquots at –20°C. The CE DNA control is thawed and analyzed at the beginning of each analytical run to assure instrument performance, based on reproducible peak fluorescence, separation, and migration times for the IS and CMV amplicons.

3.2. Quantitative Competitive PCR

Samples to be analyzed by PCR include a PCR blank containing no IS and water instead of sample DNA; the negative CMV control with 0 and 100 copies of IS DNA; both positive CMV controls with 0–10,000 copies of IS DNA; and all patient samples with 0–10,000 copies of IS DNA.

3.2.1. Plasma DNA Extraction

Careful and complete DNA extraction from plasma is essential for assay sensitivity. The following procedure was adapted from a method reported by Boom et al. *(10)* for use with plasma samples.

1. Collect 2 mL venous blood in a tube containing EDTA (separate the plasma and store up to 24 h at 4°C if the DNA cannot be extracted immediately).
2. Centrifuge the blood for 20 min at 1500*g* at room temperature.
3. Transfer 200 μL of each plasma sample to a labeled 15-mL polypropylene tube.
4. Add 1 μL carrier DNA (10.9 mg/mL, human placental DNA, Sigma) to each patient and frozen control plasma sample.
5. Add 1.8 mL lysis buffer to each sample, and mix thoroughly.
6. Add 70 μL silica suspension to each tube, and vortex for 5 s.
7. Leave the samples at room temperature (~25°C) for 10 min. Shake vigorously every minute.
8. Centrifuge the samples for 2 min at 1500*g*.
9. Use a 10-mL plastic disposable pipet to remove the supernatant. Avoid disturbing the pellet.
10. Add 1 mL wash buffer, and resuspend the silica particles with a vortex mixer.
11. Transfer the silica suspension to a labeled 1.5-mL microcentrifuge tube, and centrifuge 15 s at 10,000*g*.
12. Remove the supernatant with a disposable pipet.
13. Add 1 mL wash buffer to the reaction tube to wash the pellet, vortex well until the pellet is resuspended.
14. Centrifuge the suspension for 15 s at 10,000*g*, and remove the supernatant.
15. Repeat **steps 13–14** twice with 70% ethanol, instead of wash buffer.
16. Repeat **steps 13–14** with acetone, instead of wash buffer.
17. Place the open tube containing the silica pellet in a heating block at 56°C until dry by visual inspection (~10 min). Cover the tube with tissue to avoid aerosol contamination.
18. Resuspend the pellet by adding 200 μL deionized water preheated to 56°C, and mix with a vortex mixer.
19. Incubate the sample in a heating block for 10 min at 56°C.
20. Centrifuge the sample at 10,000*g* for 2 min at room temperature.
21. Transfer the supernatant to a new labeled 1.5-mL microcentrifuge tube.
22. Repeat steps 18–20 to make sure that the DNA is completely extracted from the silica.
23. Use vacuum centrifugation to reduce the volume of the extracted DNA solution to less than 100 μL. Determine the approximate sample volume with a pipet, then add water to bring the volume to 100 μL (*see* **Note 4**).
24. Store the extracted DNA solution at –20°C until used for PCR.

3.2.2. Competitive PCR

The authors recommend a hot-start, step-down PCR procedure to enhance amplification specificity and sensitivity. There is no apparent crossreactivity of the IE1 and IE2 primers with human DNA or DNA from other organisms. A high PCR cycle number can be used without regard to amplification linearity, because the IS and target are co-amplified and compared in multiple reactions. Using this method, the authors were able to detect CMV in six of 10 samples containing one copy of CMV DNA (this is an assessment of post-DNA extraction sensitivity, because the samples were prepared by dilution of a solution containing a known amount of CMV DNA).

1. Prepare PCR master mixes A and B as described above.
2. For each sample or control, add 10 μL extracted DNA solution to each of five labeled PCR tubes.
3. Prepare a PCR blank containing 10 μL deionized water, instead of extracted DNA solution (evidence of any amplified DNA is evidence of reagent contamination, and cause for rejection of the analysis).
4. Transfer 70 μL master mix A to each reaction tube.
5. To four of five sample reaction tubes for each sample, add 10 μL IS solution containing either 10, 100, 1000, or 10,000 copies of the IS. Add 10 μL deionized water to the fifth reaction tube and to the negative control.
6. Place three drops of mineral oil on top of each sample.
7. Close the tubes and place them in the thermocycler (Perkin-Elmer 480) with constant temperature (100°C) for 10 min, to ensure complete denaturation of the dsDNA. Remove the samples and immediately cool on ice for 10 min.
8. Add 10 μL reaction master mix B to each tube (apply to the side of the tube). Centrifuge 20 s at maximum speed in a microcentrifuge, to force the solution through the mineral oil.
9. Amplify immediately, using the following hot start/step down temperature profile. Seven steps, of five cycles each, were performed as follows (denaturation; annealing; primer extension, °C, *s*): Step 1: 94, *60;* 71, *60;* 72, *40.* Step 2: 94, *40;* 71, *60;* 72, *60.* Step 3: 94, *40;* 68, *60.* 72, *60.* Step 4: 94, *40;* 65, *60;* 72, *60.* Step 5: 94, *40;* 62, *60;* 72, *60.* Step 6: 94, *40;* 59, *60;* 72, *60.* Step 7: 94, *60;* 56, *60;* 72, *40.* The total number of cycles is 35, and the approximate time required is 2 h and 30 min.

3.2.3. Endonuclease Digestion

A large excess of *Bam*HI is used to assure complete digestion of the IS. The authors recommend that each laboratory periodically analyze a PCR sample containing the highest level of the IS in the absence of sample DNA to verify suitable enzymatic activity.

1. Transfer 20 μL of each PCR product to a labeled 0.5-mL microcentrifuge tube.
2. Add 1.25 μL (25 U) *Bam*HI restriction enzyme (20,000 U/mL, New England BioLab, Beverly, MA), 2.5 μL 10X digest buffer, 0.25 μL 100X BSA, and 1 μL deionized water to each PCR product.

3. Incubate the samples at 37°C for 4 h.
4. Remove the samples from the incubator, and analyze by CGE. Store up to 24 h at –20°C prior to analysis.

3.3. Capillary Electrophoresis

It is not necessary to analyze all five replicate patient samples, if CMV is not present. The absence of CMV is determined first by analyzing the sample that was amplified in the absence of the IS. If a CMV amplicon peak is not apparent, only the replicate sample containing 10 copies of IS DNA should be analyzed by CGE (this is to verify the presence of an IS amplicon peak, and to confirm the lack of PCR inhibitors in the patient sample). If a CMV amplicon peak is present, the remainder of the samples should be analyzed and used to calculate the number of starting copies of CMV DNA in the plasma. Since the samples without the IS do not require enzymatic digestion, they can be analyzed by CGE; the remainder of the samples are processed with the *Bam*HI.

3.3.1. Preanalytical

3.3.1.1. CAPILLARY CONDITIONING AND STORAGE

A variety of commercial capillaries, conditioning protocols, and storage conditions may be suitable for CGE. For each new capillary, follow the manufacturer's guidelines for installation on the instrument. When using sections of DB-1 cut from a bulk-purchase lot, first, manually fill the capillary with water, then prepare the detector window by heating (flame or element), to remove the external polyimide coating. Before use, rinse each capillary at high pressure with methanol, water, and electrophoresis buffer for 10 min each. The authors maintain the same rinse sequence (with shorter time duration) for daily prerun conditioning, and for between-sample conditioning. To store a capillary after each analytical run, rinse the capillary with water for 5 min, and leave both ends of the capillary in water, with the cooling system on the instrument set to 25°C. There is insufficient information on the long-term storage of capillaries off the instrument to make any useful recommendations.

3.3.1.2. INSTRUMENT SETUP

1. Configure the anode and cathode to the capillary outlet and inlet, respectively.
2. Set the temperature of the capillary cooling system to 25°C.
3. Make sure that a 488-nm rejection filter and a 520-nm band-pass filter are installed in the laser detector.
4. Turn on the laser source, and allow sufficient time for the detector output to stabilize before running samples (~15 min).
5. Set the data collection rate to 5 Hz, and rise time to 1.0.
6. Refill capillary solutions daily; or replace as needed (**Table 1**; *see* **Note 5**).

Table 1
CGE Reagents

Contents	Replace[a]	mL
Capillary Inlet		
Water (rinse and capillary storage)	Weekly	~4
Electrophoresis buffer (to fill the capillary)	Daily	0.2
Electrophoresis buffer (cathode solution)	Daily	~4
Water (to rinse the outside of the capillary)	Weekly	~4
Methanol	Weekly	~4
Capillary Outlet		
Electrophoresis buffer (anode solution)	Daily	~4
Empty waste vial	Weekly	N/A
Water (capillary storage)	Weekly	~4

aRefill as needed if not replaced. The methanol, in particular, may need more frequent filling, even during an analytical run.
N/A, not applicable.

7. Condition the capillary before each analytical run by rinsing it with methanol, water, and electrophoresis buffer for 5 min each.

3.3.1.3. SAMPLE PREPARATION

The amount of negatively charged DNA introduced into the capillary by electrokinetic injection is markedly affected by the concentration of competing charged particles in the sample. Because of the high salt concentration of the PCR and enzyme digestion reactions, the samples should be diluted with water (e.g., 100-fold) to enhance sample injection. Alternatively, the sample can be treated by membrane dialysis, as described below, to desalt the sample without dilution, thereby enhancing assay sensitivity.

1. Place a membrane filter (assure proper interface orientation) in a plastic cell-culture dish filled with deionized water.
2. Carefully pipet 25-µL digested PCR product onto each dialysis membrane (the authors have placed up to three samples on each membrane).
3. After 30 min, remove the desalted sample with a pipet and transfer to a new 0.5-mL microcentrifuge tube.
4. To dilute 100X, add 990 µL distilled water to a labeled 1.5-mL microcentrifuge tube.
5. Transfer 10 µL desalted PCR product to the tube, and mix well.
6. Transfer 25 µL diluted sample to a 0.5-mL microcentrifuge tube for analysis by CE.

3.3.2. Analytical

Configure the system so that analyses will not begin unless the capillary temperature is stable at 25°C. Make sure that all solutions and samples are in the appropriate positions on the autosampler, and are correctly identified in the instrument software. The following steps describe the various functions that should be programmed into the instrument to be performed with each sample.

1. Rinse the capillary with fresh electrophoresis buffer at high pressure (20 psi) for 1 min.
2. Inject each PCR sample (digested, desalted, and/or diluted, as necessary) at 3 kV for 10 s.
3. Move the inlet and outlet of the capillary to the cathode and anode electrophoresis buffers, respectively.
4. Ramp the voltage to 8.1 kV over 0.2 min.
5. Stop the voltage 7 min after the voltage was initiated.
6. Rinse the capillary with methanol at high pressure for 0.25 min.
7. Rinse the capillary with water at high pressure for 0.25 min.
8. Repeat **steps 1–7** for all subsequent samples.
9. At the end of each analytical run, include an equilibration method (no data collected) that will rinse the capillary with water for 5 min, and leave the capillary inlet and outlet submerged in water.

3.3.3. Postanalytical

3.3.3.1. DATA PROCESSING

With System Gold Version 8.1, the peak identification table is used to name peaks on the screen/printed output. The peak integration table delineates autointegration parameters (e.g., peak width and threshold) that will be used for peak detection and area determination.

1. In the peak identification table, enter the names of the IS and CMV target. Each day, enter the migration times observed for the CE DNA control analyzed at the beginning of the run. These will be used to identify peaks in subsequent unknown samples.
2. In the peak integration table, set the peak threshold and peak width to appropriate values (optimal values may vary; the authors have found a peak width of 0.01 and threshold of 10 to be widely applicable). It may be useful to include a minimum peak height, to suppress detection of irrelevant minor peaks.

3.3.3.2. PERFORMANCE REVIEW

The results of each control and sample should be reviewed on-screen, to verify analytical run performance and the suitability of each individual analysis.

1. Examine each electropherogram, to verify that the absorbance and current output resemble that typically expected (**Fig. 1**).
2. Verify that all peak heights are less than 1000 RFU. Detector response with the P/ACE LIF appears linear to this level, but peak area ratio calculations are inaccurate if either the IS or CMV peak height is out of range. Reanalyze the sample

with a shorter injection time, if necessary.
3. Evaluate peak autointegration on-screen for each sample, to be sure that the system has appropriately determined the beginning and end of each peak. Integration problems may be corrected manually, or by changing integration parameters.

3.4. Calculations

The authors recommend the construction of a spreadsheet program to perform the following calculations:

1. List the peak areas for the CMV and IS amplicons in each of the five replicate samples.
2. Multiply the IS peak area by 1.09, to calculate the corrected IS peak area in each replicate (*see* **Note 6**).
3. Calculate the ratio of the CMV peak area/corrected IS peak area.
4. Calculate the linear regression of the log of the CMV/IS peak area ratio determined in **step 3** (y-axis) against the log of the number of starting copies of IS DNA (10, 100, 1000, or 10,000 copies, i.e., 1, 2, 3, or 4) in the replicate samples (**Fig. 2**). Verify suitable linearity ($r^2 > 0.95$), (*see* **Note 7**).
5. Set $y = 0$ in the regression equation, and solve for x. The antilog of x is the number of starting copies of CMV DNA. The easiest way to make the calculation is shown in **Fig. 2** as $10^{-B/A}$, where **A** and **B** are the slope and intercept of the regression equation.
6. If CMV amplicons are present, but the CMV peak area is less than that of the lowest IS value, the result should be reported as less than 10 copies of CMV DNA/20L μL plasma. If the CMV peak area is greater than that of the highest IS value, the result should be reported as greater than 10,000 copies of CMV DNA/20 μL plasma.

3.5. Interpretation

PCR detection of CMV DNA is highly sensitive, but the mere presence of CMV in body tissues or fluids is not highly specific for a diagnosis of CMV disease *(8,13,14)*. This is because CMV is a ubiquitous pathogen that may be present as a primary or latent infection without pathogenicity. PCR methods for the sensitive and precise measurement of CMV viral load are relatively new and varied, such that there is little consensus information about what constitutes a normal or pathogenic infection level in different tissues. Moreover, the question of a normal or pathogenic range is somewhat moot, because that which constitutes a pathogenic viral load is highly dependent on the status of the host immune system *(15,16)*.

For this reason, a quantitative PCR assay for CMV is most useful for sensitive detection of CMV infection, and for longitudinal assessment of viral load as an indicator of changes in patient CMV status. Changes in the number of copies of CMV DNA in plasma over time appear to be a very early, presymptomatic predictor of the progression of CMV infection, allowing preemptive

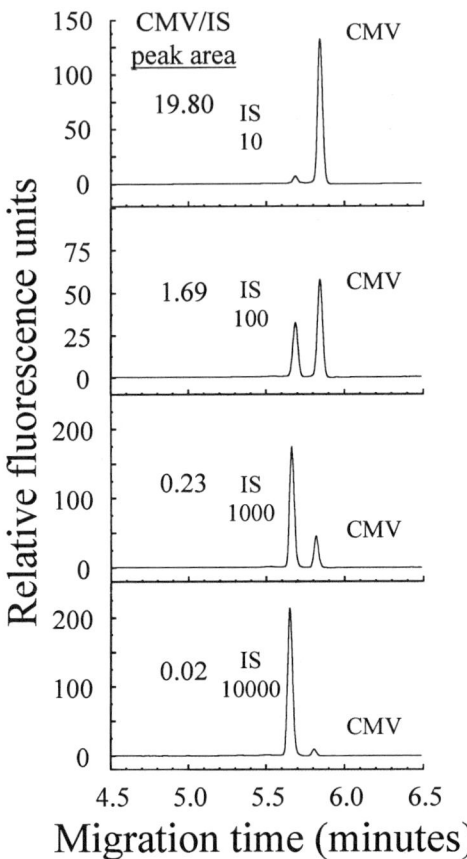

Fig. 1. Competitive PCR and CGE. The same volume of DNA solution extracted from a plasma sample was co-amplified in the presence of 10, 100, 1000, or 10,000 starting copies of IS DNA. The PCR amplicons were then separated by CGE in 0.5% HPMC in TBE containing a 1:20,000 dilution of SYBR Green I molecular dye. Baseline resolution of the enzyme digested IS (245 bp) and CMV (268 bp) amplicons was observed (primer/dimer eluted before 4.5 min). The CMV/corrected IS fluorescence peak area ratios were used to quantify the number of copies of CMV DNA in the plasma sample (**Fig. 2**).

therapeutic intervention. Similarly, longitudinal assessment of CMV load can be used to track the response to therapy, allowing more confident evaluation of patient status, and earlier cessation of antibiotics (*14,16,17*).

3.6. Summary

The method described in this chapter for the quantification of CMV DNA combines the specificity and sensitivity of viral DNA amplification by com-

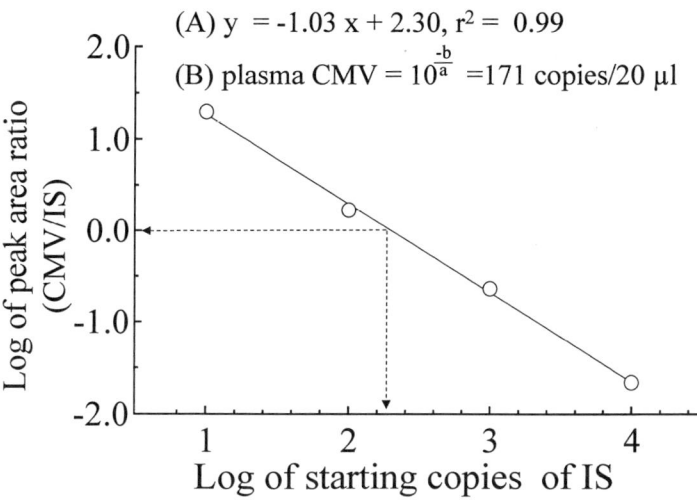

Fig. 2. Linear regression analysis of competitive PCR results. A log/log standard curve prepared from the results in **Fig. 1** showed excellent response linearity. The slope and intercept of the regression (**Eq. A**) indicate the point on the line where the peak area ratios are equal (log 1 = 0). These numbers are then used in **Eq. (B)** to calculate the number of copies of CMV DNA in the plasma sample. The value is expressed per 20 µL plasma, because 10 µL extracted DNA solution was used in the PCR reaction (DNA was extracted from 200 µL of plasma into 100 µL DNA solution).

petitive PCR with the sensitive, automated detection of DNA amplicons by CGE. New information arising from the more widespread use of this and other quantitative PCR methods for assessment of CMV viral load should rapidly add to the limited knowledge of the transmission and progression of CMV disease. A particularly important application of this test is the correlation of CMV load and changes in CMV load in plasma, peripheral blood leukocytes, and other tissues as prognosticators of CMV disease in a host of at-risk patients, especially neonates, organ transplant recipients, and other immunocompromised individuals.

4. Notes

1. The IS was prepared by PCR site-directed mutagenesis from DNA extracted from cultured CMV. The forward primer was the IE-1 sequence described above. The reverse primer sequence was 5' GGT GCT CAC GCA CAT <u>GGA TCC</u> CAT A 3'. The reverse primer sequence introduces a *Bam*HI restriction site (underlined) into DNA amplified from the wild type CMV sequence. The mutated PCR product was then cloned into plasmid DNA, using a TA cloning kit (Invitrogen, San Diego, CA). To prepare the IS solutions for competitive PCR, an aliquot of the

stock plasmid DNA solution was first linearized by *Hind*III enzymatic digestion. The DNA concentration of this solution was then determined by absorbance at 260 nm, and diluted into four solutions containing 10, 100, 1000, or 10,000 copies of IS DNA/10 μL. For routine use, the authors make 2 mL (enough for 200 samples) of each solution, and store them in 100-μL aliquots at –20°C.

With this method of preparing the IS, similar amplification efficiency of the IS and CMV target is expected, because the sequences are similar in size and composition. A disadvantage of this method is the time required for enzymatic digestion. An attractive alternative is the use of a loop primer to make an IS producing a PCR product that is slightly larger than the CMV target *(18)*.

2. A higher-than-standard MgCl$_2$ concentration (5 mM) is needed for optimal amplification of DNA in blood samples collected using EDTA as an anticoagulant.

3. Conditioned with methanol before initial use, before each daily run, and between samples, a single DB-1 capillary (~$10, if purchased in bulk) will usually give reproducible results for several hundred samples. Capillary suitability or failure is best detected by a loss of quantification accuracy when evaluating the controls.

4. Reconstitution of a completely dry DNA pellet is difficult, and may result in poor recovery, if attempted too soon. The authors have observed that closely approximating the reconstituted volume of the extracted DNA solution, as described, is more efficient and practical in routine application.

5. For the Beckman P/ACE, water-filled vials should be kept in tray positions 1 and 11, because these are the default capillary docking positions in the event of a system power failure. It is therefore practical to also use these vials as the docking sites for the capillary when the instrument is not in use.

6. The fluorescence peak area of PCR amplicons is directly proportional to DNA mass concentration, rather than to molar concentration. Competitive PCR, however, compares the molar amounts of the IS and CMV amplicons. As a consequence, the peak area of the IS amplicons must be multiplied by a correction factor (1.09, the ratio of the number of base pairs in each amplicon, or 268/245) to normalize the peak areas to a molar equivalent.

7. The linearity of the log/log response curve is typically excellent, unless there is great disparity between the starting copies of the IS and CMV DNA. If the linearity of the response using all four replicate samples containing the IS is not acceptable ($r^2 < 0.95$), the end point furthest from the observed CMV:corrected IS peak area ratio can be omitted from the regression analysis.

References

1. Nelson, C. T., Istas, A. S., Wilkerson, M. K., and Demmler, G. J. (1995) PCR detection of cytomegalovirus DNA in serum as a diagnostic test for congenital cytomegalovirus infection. *J. Clin. Microbiol.* **33,** 3317–3318.

2. Gerna, O., Furione, M., Baldanti, F., Percivalle, E., Comoli, P., and Locatelli, F. (1995) Quantitation of human cytomegalovirus DNA in bone marrow transplant recipients. *Br. J. Haematol.* **91,** 674–683.

3. Wolff, C., Skourtopoulos, M., Hornschemeyer, D., Wolff, D., Korner, M., Huffert, F., Korfer, R., and Kleesier, K. (1996) Significance of human cytomegalovirus DNA detection in immunocompromised heart transplant patients. *Transplantation* **61,** 750–757.

4. Tanabe, K., Takahashi, K., Koyama, I., et al. (1996) Early diagnosis of CMV syndrome after kidney transplantation: comparison between CMV antigenemia and PCR assay. *Transplant Proc.* **28,** 1508–1510.

5. Arribas, R., Clifford, D. B., Fichtenbaum, C. J., Commins, D. L., Powderly, W. G., and Storch, G. A. (1995) Level of cytomegalovirus (CMV) DNA in cerebrospinal fluid of subjects with AIDS and CMV infection of the central nervous system. *J. Infect. Dis.* **172,** 527–531.

6. Shinkal, M., Bozzette, S. A., Powderly, W., Frame, P. ,and Spector, S. A. (1997) Utility of urine and leukocyte cultures and plasma DNA polymerase chain reaction for identification of AIDS patients at risk for developing human cytomegalovirus disease. *J. Infect. Dis.* **175,** 302–308.

7. Myers, J. B. and Amsterdam, D. (1997) Laboratory diagnosis of cytomegalovirus infections. *Immunol. Invest.* **26,** 383–394.

8. Imbert-Marcille, B. M., Cantarovich, D., Ferre-Aubineau, V., Richet, B., Soulillou, J. P., and Billaudel. (1997) Usefulness of DNA viral load quantification for cytomegalovirus disease monitoring in renal and pancreas/renal transplant recipients. *Transplantation* **63,** 1476–1481.

9. Kotsimbos, A. T., Sinickas, V., Glare, E. M., Esmore, D. S., Snell, G. I., Walters, E. H., and Williams, T. J. (1997) Quantitative detection of human cytomegalovirus DNA in lung transplant recipients. *Am. J. Respir. Crit. Care Med.* **156,** 1241–1246.

10. Boom, R., Sol, C. J., Salimans, M. M., Jansen, C. L., Wertheim-van, Dillen, P. M., and van der Noordaa. (1990) Rapid and simple method for purification of nucleic acids. *J. Clin. Microbiol.* **28,** 495–503.

11. Zipeto, D., Baldanti, F., Zella, D., Furione, M., Cavicchini, A., Milanesi, G., and Gerna, G. (1993) Quantification of human cytomegalovirus DNA in peripheral blood polymorphonuclear leukocytes of immunocompromised patients by the polymerase chain reaction. *J. Virol. Methods* **44,** 45–55.

12. Skeidsvoll and Ueland, P. M. (1995) Analysis of double-stranded DNA by capillary electrophoresis with laser-induced fluorescence detection using the monomeric dye SYBR green I. *Anal. Biochem.* **231,** 359–365.

13. Lao, W. C., Lee, D., Burroughs, A. K., Lanzini, G., Rolles, K., Emery, V. C., and Griffiths, P. D. (1997) Use of polymerase chain reaction to provide prognostic information on human cytomegalovirus disease after liver transplantation. *J. Med.Virol.* **51,** 152–158.

14. Kusne, S., Manez, R., Frye, B. L., St. George, K., Abu-Elmagd, K., Tabasco-Menguillon, J., et al. (1997) Use of DNA amplification for diagnosis of cytomegalovirus enteritis after intestinal transplantation. *Gastroenterology* **112,** 1121–1128.

15. Cope, A. V., Sabin, C., Burroughs, A., Rolles, K., Griffiths, P. D., and Emery, V. C. (1997) Interrelationships among quantity of human cytomegalovirus (HCMV) DNA in blood, donor-recipient serostatus, and administration of methylpredniso-

lone as risk factors for HCMV disease following liver transplantation. *J. Infect. Dis.* **176,** 1484–1490.

16. Toyoda, M., Carlos, Galera, O. A., Galfayan, K., Zhang, X., Sun, Z., Czer, L. S., and Jordan, S. C. (1997) Correlation of cytomegalovirus DNA levels with response to antiviral therapy in cardiac and renal allograft recipients. *Transplantation* **63,** 957–963.

17. Brytting, M., Mousaviazi, M., Bostrom, L., Larsson, M., Lunderberg, J., Ljungman, P., Ringden, O., and Sundqvist, V. A. (1995) Cytomegalovirus DNA in peripheral blood leukocytes and plasma from bone marrow transplant recipients. *Transplantation* **60,** 961–965.

18. Poirier-Toulemonde, A. S., Imbert-Marcille, B. M., Ferre-Aubineau, V., Besse, B., LeRoux, M. G., Cantarovich, D., and Billaudel. (1997) Successful quantification of cytomegalovirus DNA by competitive PCR and detection with capillary electrophoresis. *Mol. Cell. Probes* **11,** 11–23.

10

Laboratory Diagnosis of Structural Hemoglobinopathies and Thalassemias by Capillary Isoelectric Focusing

James M. Hempe and Randall D. Craver

1. Introduction

Structural hemoglobinopathies and thalassemias are congenital hemoglobin (Hb) disorders that cause anemia, morbidity, and mortality resulting from abnormal Hb function. Structurally different normal Hb variants include HbA_2, HbF (fetal hemoglobin), and HbA (adult hemoglobin). Each is a protein tetramer consisting of two α-globins, and either two δ–, γ-, or β-globins, respectively. Fetal Hb ($\alpha_2\gamma_2$) predominates in neonates (60–95% of total Hb), but declines to <1% in older children and adults. HbA_2 ($\alpha_2\delta_2$) is a minor constituent with apparently normal function that is not detected in neonates, but increases to about 2–3% of HbA ($\alpha_2\beta_2$) in older children and adults. Mutations in the genes that regulate the structure and synthesis of α-, β-, γ- and δ-globins produce abnormal and often dysfunctional Hb variants (structural hemoglobinopathy), or cause decreased synthesis of normal Hb variants (thalassemia) (1,2). DNA mutations that cause structural hemoglobinopathies are most commonly diagnosed by indirect assays that identify abnormal gene products (i.e., Hb variants) rather than abnormal genes. Over 600 abnormal structural Hb variants have been reported (3), most of which (95%) differ from normal HbA by replacement of a single amino acid (2). Although some structural mutations are benign, many (50% of β-variants and 20% of α-variants) alter Hb solubility, stability, or oxygen affinity in ways that adversely affect Hb function. In contrast, thalassemia syndromes are caused either by deletions of entire genes or by mutations that affect the production or processing of normal globin mRNAs. Abnormal Hb variants are not produced in the thalassemias, and thus cannot be used to diagnose the underlying mutation. Nevertheless, Hb variant

From: *Methods in Molecular Medicine, Vol 27: Clinical Applications of Capillary Electrophoresis*
Edited by: S. M. Palfrey © Humana Press Inc., Totowa, NJ

analysis provides diagnostic information about thalassemia mutations, because the relative proportions of normal Hb variants, especially HbA_2 and HbF, are often disturbed when the synthesis of α-, β-, or γ-globin is abnormal. Also, abnormal globin products that are not allelic Hb variants are detected in patients with HbH disease, a severe form of α-thalassemia in which the lack of α-globin production leads to the association of excess β- or γ-globins into unstable tetramers called HbH or Bart's, respectively.

1.1. Laboratory Analysis

An extensive variety of genotypes and phenotypes, including those caused by multiple gene mutations, comprise this class of congenital Hb disorders. It is not surprising, therefore, that no single analytical test can cost-effectively diagnose all possible disease states. In fact, comprehensive laboratory evaluation of some of the more unusual genetic disorders can require a series of progressively more esoteric and costly tests to confirm a diagnosis. Although extensive confirmatory testing is usually relegated to highly specialized and qualified reference laboratories, primary testing during preliminary evaluation of a suspected congenital hemoglobin disorder remains primarily the purview of the community hospital or regional reference lab. The most commonly used conventional diagnostic tests include alkaline electrophoresis, acid electrophoresis, minicolumn ion exchange chromatography, and alkali denaturation. Individually, each test lacks sensitivity and specificity for the detection and diagnosis of both structural hemoglobinopathies and thalassemias. Collectively, although the tests provide adequate information for preliminary evaluation of hemoglobin disorders, the assay procedures are labor intensive and unsuitable for automation. High-performance liquid chromotography (HPLC) is an automated analytical method, but more expensive and less frequently used for primary testing. This chapter describes the use and advantages of automated capillary isoelectric focusing (IEF) as a comprehensive primary clinical test for the evaluation of structural hemoglobinopathies and thalassemia syndromes.

1.2. Overview of Capillary Isoelectric Focusing

Human hemoglobin variants are amphoteric molecules that can be separated as protein dimers (2), based on surface charge by isoelectric focusing over a linear range of 6–8 pH units. In the cIEF method described here (**Fig. 1**), a 50 μm id × 27 cm long DB-1 capillary is first filled with carrier ampholytes in liquid polymer solution. Next, pressure is used to inject a small volume of hemolyzed blood into the capillary inlet, which is then immersed in dilute phosphoric acid solution at the anode. The capillary outlet is simultaneously immersed in dilute sodium hydroxide solution at the cathode, and 30 kV is

Injection

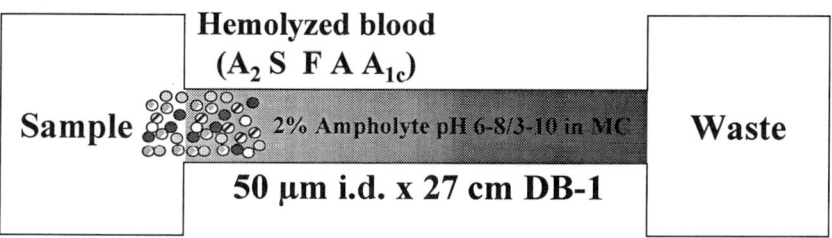

Separation

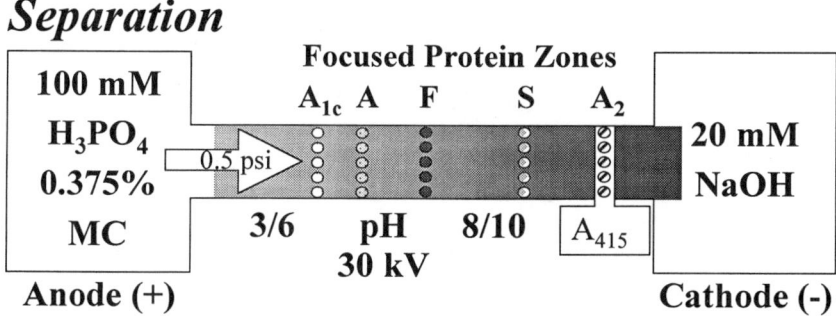

Fig. 1. Principle of cIEF. Hemolyzed blood from a patient with sickle cell trait is pressure-injected at the anode end of a capillary filled with ampholyte–methylcellulose (MC) solution. When voltage is applied, the proteins migrate to their respective isoelectric points (pIs). Low pressure at the anode is applied to elute the focused proteins past the detector.

applied across the electrodes, generating an electric field of 1111 V/cm. Current through the system increases rapidly to ~4 µA then subsides to ~1 µA within 2–3 min, as the ampholytes distribute in different zones inside the capillary (**Fig. 2**). A linear pH gradient is rapidly established, with the low-pH region adjacent to the anolyte. While the sample is located between the phosphoric acid and the low end of the ampholyte pH gradient, the net surface charge of each Hb variant is positive. The variants are thus repulsed by the anode and attracted to the cathode. As the Hb variants traverse regions of progressively higher pH, positively charged sites on their surfaces become deprotonated or hydroxylated. Migration slows as the surface charge decreases toward neutrality. Eventually, each different Hb variant reaches a pH region where the positive and negative charges on the surface of the molecule are balanced, and migration stops. The molecules of each Hb variant become concentrated in a narrow band at the pH that is the isoelectric point (pI) of that particular protein.

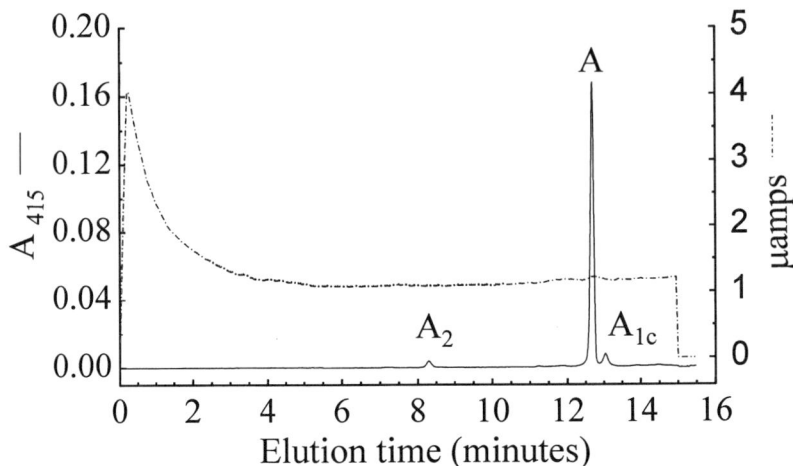

Fig. 2. Representative cIEF electropherogram. The Hb variants normally present in blood are HbA$_2$, -A, and -A$_1$c. Quantification based on integration of absorbance peak area is precise for all variants, despite the large differences in their concentrations. The shape of the current profile is characteristic; constant current indicates the completion of ampholyte distribution inside the capillary.

The process is called isoelectric focusing, because any of the concentrated molecules that diffuse away from the pI into regions of lower or higher pH will gain or lose electrons, respectively. When this happens, the charged molecule will be attracted/repulsed from the two electrodes until it again returns to its pI, and again becomes neutral. After the Hb variants focus in this manner, pressure (0.5 psi) is generated at the inlet of the capillary to mobilize the proteins; the voltage is still applied to keep each band from diffusing. The focused protein zones are thus pushed past the detector window, where absorbance at 415 nm (specific for heme moieties) is measured. Detector output is collected by a computer and displayed on-screen as a series of absorbance peaks. **Figure 2** is a full-scale electropherogram showing HbA$_2$, -A, and -A$_1$c (a glycosylated postranslational modification of HbA) in a normal sample. **Figure 3** shows an electropherogram obtained by analysis of blood from a hypertransfused patient with HbS/C disease containing HbC, -A$_2$, -S, -F, -A, and -A$_1$c. Data analysis software is used to integrate the absorbance output, to determine the area under the curve above baseline for each Hb peak. The amount of each variant present is then expressed as a proportion (% of total Hb), i.e., the area of each peak divided by the total area of all peaks. Since most Hb variants have different pIs, they can be identified by where they focus in the capillary. Consequently, peak identification is based on the calculated pIs of unknown peaks, deter-

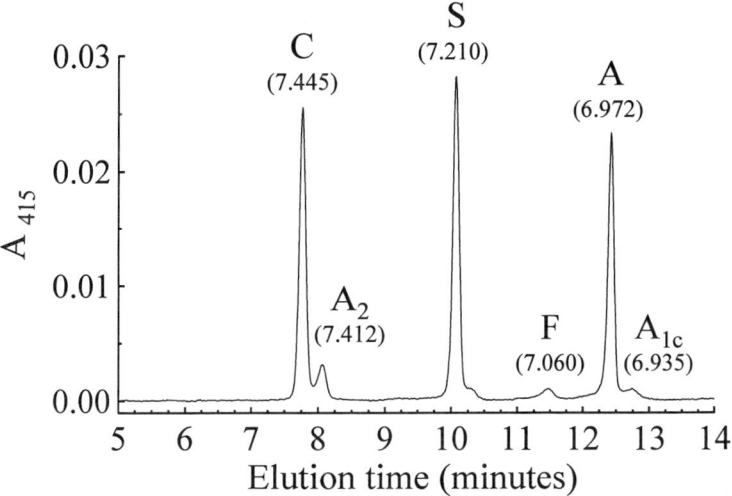

Fig. 3. Common abnormal Hb variants. Blood from a patient with HbS/C disease undergoing hypertransfusion was analyzed by cIEF. The electropherogram shows the normal HbA$_2$, -F, -A, and -A$_1$c, and HbsS and C, two of the most common abnormal variants in the world. The pI of Hb variants in a patient sample are determined by linear regression of elution time against the pI of known Hb variants in a standard. The isoelectric points of all known Hb variants lie between 6 and 8 pH units.

mined by linear regression of pIs against elution time of known Hb variants in a standard sample.

2. Materials

Analytical-grade reagents should be used unless otherwise noted. Suitable performance of all new lots of reagents and materials should be verified before reporting clinical results. At a minimum, the proportions of the different Hb variants in the normal and abnormal controls should be within expected tolerance limits previously defined by replicate analysis.

1. DB-1 capillary: J&W Scientific (Folsom, CA), 50 μm id, 0.2 μm coating thickness (*see* **Note 1**).
2. Methylcellulose solution: 0.375% methylcellulose (1500 cp at 2%, Sigma, St. Louis, MO) in deionized water. Add 250 mL deionized water to a 1 L glass bottle, and heat to 65 ± 5°C (~90 s in a microwave oven). Add 0.9275 g methylcellulose, place the bottle on a stir plate, and mix for 10 min. Partially submerge the bottle in a container of ice water and mix on a stir plate until the solution reaches 5–10°C (~30 min). Remove the bottle from the ice water (the solution should be clear and free of particles), and mix until the solution reaches room temperature (~25°C, 30 min). Vacuum-filter (0.45 μm, more than one filter may be required) and store up to 6 mo at 4°C.

3. 1 M NaOH solution: Add 4 g NaOH to a 100-mL volumetric flask, fill to the mark with deionized water, and mix thoroughly. Store indefinitely in a plastic bottle at room temperature.

4. 5 M H_3PO_4 solution: Add 11.25 mL stock phosphoric acid solution *(85%)* to a 100-mL volumetric flask, fill to the mark with deionized water, and mix thoroughly. Store indefinitely in a plastic bottle at room temperature.

5. Stock mixed-ampholyte solution (10:1): Transfer 1.0 mL pH 6.0–8.0 ampholyte and 0.1 mL pH 3.0–10.0 ampholyte (Pharmalyte, Sigma) to an amber screw-capped vial, mix thoroughly, and store indefinitely at 4°C.

6. Ampholyte–methylcellulose solution: Transfer 4.9 mL methylcellulose solution to an amber vessel. Add 0.1 mL (2%) of the stock mixed-ampholyte solution. Mix thoroughly, and store for up to 1 mo at 4°C. Let stand at least overnight prior to use (*see* **Note 2**)

7. Cathode solution: 20 mM NaOH in deionized water. Add 4.0 mL 1 M NaOH solution to a 200-mL volumetric flask. Fill to the mark with deionized water, and mix thoroughly. Store indefinitely at room temperature in a plastic squeeze bottle with a flip-top spout.

8. Anode solution: 100 mM H_3PO_4 in methylcellulose solution. Add 4.0 mL 5 M H_3PO_4 to a 200-mL volumetric flask. Fill to the mark with methylcellulose solution, and mix thoroughly. Store up to 6 mo at 4°C in a plastic squeeze bottle with a flip-top spout.

9. Hemolyzing reagent: 5 mM EDTA and 10 mM KCN in deionized water. In a fume hood, add 0.465 g EDTA disodium salt and 0.1625 g KCN (Aldrich, Milwaukee, WI) to a 250-mL volumetric flask. Fill to the mark with deionized water, mix, vacuum filter (0.45 μm), and store tightly capped up to 6 mo at room temperature in an amber bottle.

10. A dog nail clipper from a pet store is invaluable for converting 0.5-mL microcentrifuge tubes into topless sample vials.

3. Methods

The described method is based on previously published research from this laboratory *(4–7)*. Some degree of familiarity with capillary electrophoresis (CE) instrumentation and operation is assumed. Procedural steps are specifically described, as used with a P/ACE 2200 CE instrument with version 8.1 of System Gold Software (Beckman, Fullerton, CA). Adaptation of these instructions to other platforms should be straightforward. Before reporting clinical results, each laboratory should validate their method by demonstrating suitable sensitivity, specificity, precision, linearity, and correlation with accepted diagnostic methods.

3.1. Preparation of Unassayed Controls

Normal and abnormal control samples should be evaluated with each analytical run to verify instrument performance for both quantification and identification. The authors recommend the use of frozen liquid red blood cell (RBC) hemolysates, because commercially available lyophilized controls contain hemoglobin oxidation products (e.g., methemoglobins) that adversely affect assay precision and sensitivity. Normal and abnormal controls should be prepared by mixing RBC with hemolyzing reagent in the same 1:20 ratio used to prepare patient samples (*see* **Subheading 3.2.1.3.**). The hemolysates can be stored in 200-µL aliquots in 0.5 mL microcentrifuge tubes at –70°C (storage at –20°C is not acceptable) for well over 1 yr, with no effect on the measured proportions of HbA$_2$, -S, -F, or -A. Blood used to prepare the normal control should contain no measurable HbF, and normal levels of HbA$_2$ and -A. The abnormal control should be blood from a patient with S/β$^+$-thalassemia (or pooled blood mixed to reflect a similar Hb composition) containing HbA$_2$, -S, -F, and -A, with elevated amounts of HbA$_2$ (~5%) and HbF (5–10%). HbC should not be included in controls used to monitor Hb quantification, because it interferes with the measurement of HbA$_2$.

3.2. Capillary Isoelectric Focusing

3.2.1. Preanalytical

3.2.1.1. CAPILLARY CONDITIONING AND STORAGE

A variety of commercial capillaries, conditioning protocols, and storage conditions may be suitable for cIEF. The authors recommend the use of DB-1 capillaries, for reasons previously described (*6*). For each new capillary, follow the manufacturer's guidelines for installation on the instrument. When using sections of DB-1 cut from a bulk purchase lot, first manually fill the capillary with water, then prepare the detector window by briefly heating (flame or element) to remove the external polyimide coating. Before use, rinse each capillary at high pressure with methanol, water, and ampholyte solution for 10 min each. The authors maintain the same rinse sequence (with shorter time duration) for daily prerun conditioning and for between-sample conditioning. To store a capillary after each analytical run, rinse the capillary with water for 5 min, and leave both ends of the capillary in water, with the cooling system on the instrument set to 20°C. There is insufficient information on the long-term storage of used capillaries off the instrument to make any useful recommendations.

3.2.1.2. INSTRUMENT SETUP

1. Configure the anode and cathode to the capillary inlet and outlet, respectively.
2. Set the temperature of the capillary cooling system to 20°C.
3. Make sure that a 415 nm band-pass filter is installed and in position (*see* **Note 3**).
4. Turn on the lamp and allow sufficient time for the detector output to stabilize before running samples (~15 min).
5. Set the data collection rate to 5 Hz, and rise time to 1.0.

Table 1
cIEF Reagents

Contents	Replace[a]	mL
Capillary Inlet		
Water (rinse and capillary storage)	Weekly	~4
Ampholyte/methylcellulose solution	Daily	0.2
Anode solution (H_3PO_4)	Daily	~4
Methanol	Weekly	~4
Capillary Outlet		
Cathode solution (NaOH)	Daily	~4
Empty waste vial	Weekly	N/A
Water (capillary storage)	Weekly	~4

[a]Refill as needed when not replaced.

6. Refill capillary solutions daily, or replace as needed (**Table 1**; *see* **Note 4**). At least 30 samples can be analyzed without changing solutions.
7. Condition the capillary before each analytical run by rinsing it with methanol, water, and ampholyte–methylcellulose solution for 5 min each.

3.2.1.3. SAMPLE PREPARATION

The authors use blood collected in EDTA, but have observed no differences in samples collected with heparin as the anticoagulant. Quantitative results from blood stored for short periods (several hours) at room temperature, or up to a week at 4°C, are similar to those observed in fresh blood.

1. Add 200 μL hemolyzing reagent to labeled 0.5-mL centrifuge tubes.
2. Transfer 0.5 mL whole blood from each sample to a labeled 1.5-mL centrifuge tube (this is only a convenient volume, much less can be used).
3. Centrifuge at 1000*g* for 5 min at room temperature to separate plasma and RBC.
4. Pass a pipet tip through the plasma layer and transfer 10 μL RBC to the hemolyzing reagent (*see* **Note 5**). Thoroughly rinse the inside of the pipet tip by up and down pipeting of the hemolysate.
5. Vortex briefly to enhance hemolysis and sample homogeneity.
6. Place the sample in the appropriate vial position on the autosampler.

3.2.2. Analytical

Configure the system so that analyses will not begin unless the capillary temperature is stable at 20°C. Make sure that all solutions and samples are in the appropriate positions on the autosampler, and are correctly identified in the instrument software. The authors have found it most convenient to store instrument output in subdirectories by year/month/day analyzed, and by using the laboratory sample accession number for identification. The authors strongly recommend analysis of both a normal and abnormal control at the beginning of each analytical run, to assure proper instrument operation. The following steps describe the various functions that should be programmed into the instrument to be performed with each sample.

1. Rinse the capillary with ampholyte–methylcellulose solution at high pressure (20 psi) for 1 min.
2. Inject sample hemolysate at low pressure (0.5 psi) for 10 s (*see* **Note 6**).
3. Move the inlet and outlet of the capillary to the anode and cathode solutions, respectively.
4. Ramp the voltage to 30 kV over 0.2 min.
5. Allow the proteins to focus by holding the voltage constant for 5 min prior to mobilization.
6. At 5 min, begin a low-pressure mobilization toward the outlet for 10 min. The ends of the capillary should remain in the electrode solutions, with the voltage applied during this step. The anode solution contains methylcellulose, so that the viscosity of the solution will be similar to that of the ampholyte solution, and the mobilization rate will remain constant during elution.
7. Stop the voltage and pressure 15 min after the voltage is initiated.
8. Rinse the capillary with methanol at high pressure for 1 min. Continue to collect detector output for 0.5 min during the methanol rinse (*see* **Note 7**).
9. Rinse the capillary with water at high pressure for 0.5 min.
10. Repeat **steps 1–9** for all subsequent samples.
11. At the end of each analytical run, include an equilibration method (no data collected) that will rinse the capillary with water for 5 min; leave the capillary inlet and outlet submerged in water, and turn off the lamp if the instrument is not to be used again for over 5 h.

3.2.3. Postanalytical

3.2.3.1. DATA PROCESSING

The extent of the data reduction possible with commercial CE instrumentation is largely dependent on the software provided by the manufacturer. With System Gold Version 8.1, the peak identification table is used to name Hb variants on the screen/printed output and to assign reference peaks for the calculation of corrected elution times. The peak integration table delineates autointegration parameters (e.g., peak width and threshold) that will be used for peak detection and area determination. The mol wt table is used to calculate the pI of unknown peaks by linear regression of elution time against the pI of known Hb variants in a standard.

1. In the peak identification table, enter the names of HbA_2, -S, -F, -A, and -A_1c and their elution times, observed in the abnormal control analyzed at the beginning of

the analytical run. Assign HbA as a reference peak to be used in the calculation of corrected elution times.

2. In the peak integration table, set peak threshold to 0.004 for the entire analysis (lower values will integrate irrelevant and nondiagnostic posttranslational products). Set peak width to 0.2 from 0 min, until after the elution time of HbA_2, to 0.8 or wider, covering the region surrounding the elution time of HbS (to include partially resolved glycosylated HbS in the calculation of total HbS), and to 0.25 prior to the elution time of Hb F through the end of the analysis. It may be useful to include a minimum peak height, to suppress detection of nondiagnostic minor peaks (*see* **Note 8**).

3. In the mol wt table, enter the names and pI of HbA_2, -S, -F, -A, and -A_1c (7.412, 7.210, 7.060, 6.972, and 6.935, respectively. It may be necessary to omit the decimal point in lower versions of System Gold). Also, enter the elution times of each variant from the analysis of the abnormal control at the beginning of the run. Under these conditions, the abnormal control functions as a pI standard for the current analytical run. The pI of peaks observed in subsequent samples will be based on the regression equation determined by the analysis of the control. Use corrected elution times to calculate pI to minimize the effects of minor between-run and within-run elution time variability on peak identification.

4. In the report format, include area percent as part of the screen/printed output. Area percent expresses the area of an individual peak as a proportion of total peak area. This is the value (% of total Hb) that should be reported.

3.2.3.2. Performance Review

The results of each control and sample analyzed must be reviewed to verify analytical run suitability and the suitability of each individual analysis. This is most easily accomplished on-screen.

1. Examine each electropherogram to verify that the absorbance and current output resemble that typically expected (**Fig. 2**). The presence of methemoglobins (pI 7.295, 7.162, 7.131) or HbA_3 (aging band, pI 6.854) are evidence of poor sample storage, and may be grounds for sample rejection (*6*). Anomalous minor peaks can sometimes be removed by manual integration, without affecting the interpretation. Reanalyze the sample if there is evidence of significant anomalous spikes, excessive baseline noise, or an unusual shift in elution times.

2. Verify that all peak heights are less than 0.2 absorbance units (AU). Detector response with the P/ACE 2200 is only linear to this absorbance, and HbA_2 levels will be falsely elevated if the peak height of HbA or another major variant exceeds this limit. If any peak height is greater than 0.2 AU, reanalyze the sample with a shorter sample injection time.

3. Evaluate peak autointegration on-screen for each sample to be sure that the system has appropriately determined the beginning and end of each peak. Minor integration problems may be corrected manually, or by temporarily changing integration parameters (*see* **Note 8**).

4. For routine identification of common normal and abnormal Hb variants, verify that the observed pI is equal to the expected pI ±0.01 pH units. As described earlier, the authors recommend the use of HbA as a reference peak for the determination of pI, based on corrected elution time. If HbA is not present, as in patients with sickle cell disease, the pI of the observed peaks will be nonsensical. In this case, verify that the major peak in the unknown sample is HbS by comparing the elution time to that of HbS in the abnormal control; overlaying the electropherograms of both the unknown sample and abnormal control on-screen to compare focusing positions; and making HbS the reference peak, and recalculating the pI (the pI for HbS, -A_2, and -F should now be equal to the expected pI +0.01 pH units). If the calculated pI for a peak is not recognizable as that of a common normal or abnormal variant, an uncommon abnormal variant may be present, and further analysis may be required (see **Subheading 3.3.3.**).
5. Check the absorbance output collected during the methanol rinse for the presence of an anomalous peak. Reanalyze the sample with a longer elution time, if a significant unusual peak is observed in the rinse (*see* **Note 7**).

3.3. Interpretation

The human genome includes one gene pair each for β- and δ-globin, and two gene pairs each for α- and γ-globin *(2)*. Hemoglobin variants that are the products of these genes are inherited as co-dominant traits, according to classic Mendelian genetics. As a consequence, an individual may be homozygous for normal or abnormal alleles of a specific globin gene (e.g., only wild-type $β^A$ in normal individuals, or only $β^S$ in sickle cell disease) or heterozygous (e.g., both $β^A$ and $β^S$ alleles in sickle cell trait). Double heterozygotes are also encountered when an individual possesses two abnormal alleles of the same gene (e.g., $β^S$ and $β^C$ in Hb S/C disease), and therefore produces two different abnormal Hb variants. Because all hemoglobins are made up of two different globins, an individual may express two abnormal structural alleles for different globin genes (e.g., $β^S$ and $α^G$ in Hb S/G disease), producing four different Hb variants by cIEF (or six in neonates producing γ-globin and Hb F). Compound heterozygotes can also have both a structural allele and a thalassemia allele of the same gene or different genes in which both an abnormal Hb variant and low production of a normal variant are observed (e.g., $β^S$ and $β^+$ in Hb S/$β^+$ thalassemia).

Clearly, laboratory tests used for the initial evaluation of patients with suspected congenital hemoglobin disorders must be highly sensitive and able to detect a plethora of complex genetic disorders. Hb variant analysis is most often requested after clinical observations, family history, and/or standard hematology tests (e.g., a complete blood count and blood smear) have indicated the possibility of a congenital hemoglobin disorder. Serum ferritin or other ancillary tests may also have been conducted to test for iron deficiency as a cause of anemia. The role of a primary assay for Hb variant analysis is to quickly and inexpensively identify individuals with structural hemoglobinopathies or thalassemias, even if the specific underlying mutation cannot be immediately identified. The need for expensive confirmatory testing, however, is reduced, if the primary assay also provides specific information that permits confident diagnosis of the genetic defect. Capillary isoelectric focusing can detect most structurally abnormal Hb variants, and can uniquely identify and quantify many normal and abnormal variants. The chief advantage of cIEF is that it economically combines both the sensitivity needed to screen patient samples and the specificity to confirm many diagnoses in a single automated assay.

3.3.1. Diagnostic Strategy with CIEF

Figure 4 is a simplified flowchart that uses a series of binary questions to help interpret cIEF results. Three important pathways are emphasized: Abnormal levels of normal variants, i.e., HbA_2 and -F, may indicate a thalassemia mutation; the presence of abnormal major and minor variants in the same individual is usually indicative of a structural α-globin mutation; and the presence of an abnormal major variant in the absence of an abnormal minor variant is usually indicative of a structural β-globin mutation. It should be remembered that the flowchart is an oversimplification of the interpretive process, and is most useful for the diagnosis of common disorders. Interpretations can differ greatly for double heterozygotes, patients with rare hemoglobin disorders, or infants with high levels of HbF. Clinical phenotype should always be considered in the interpretation of laboratory results. When in doubt, family studies or more definitive genetic analyses may be necessary.

3.3.2. Identifying and Reporting Common Hb Variants

HbS, -E, and -C, respectively, are the three most common structural hemoglobin variants in the world *(2)*. HbS and -C, but not HbE, are identified with high probability by cIEF, based on calculated pI without a confirmatory test. In this laboratory, cIEF is routinely used without further testing to specifically diagnose sickle cell trait, sickle cell disease, HbC trait, HbC disease, HbS/C disease, heterozygous β-thalassemia, HbS/β+-thalassemia, and hereditary persistence of fetal hemoglobin (HPFH), among others. Specific recommendations for reporting the results include:

1. Report HbC, -A_2, -S, -F, and -A without further evaluation, if the observed quantity is within the expected range, and if the calculated pI is within ±0.01 pH units of the expected pI (7.445, 7.412, 7.210, 7.060, and 6.972, respectively).
2. To make the data resemble that reported by conventional methods, the proportions of HbA and -A_1c (pI 6.935) should be combined to compute total HbA in the final report.

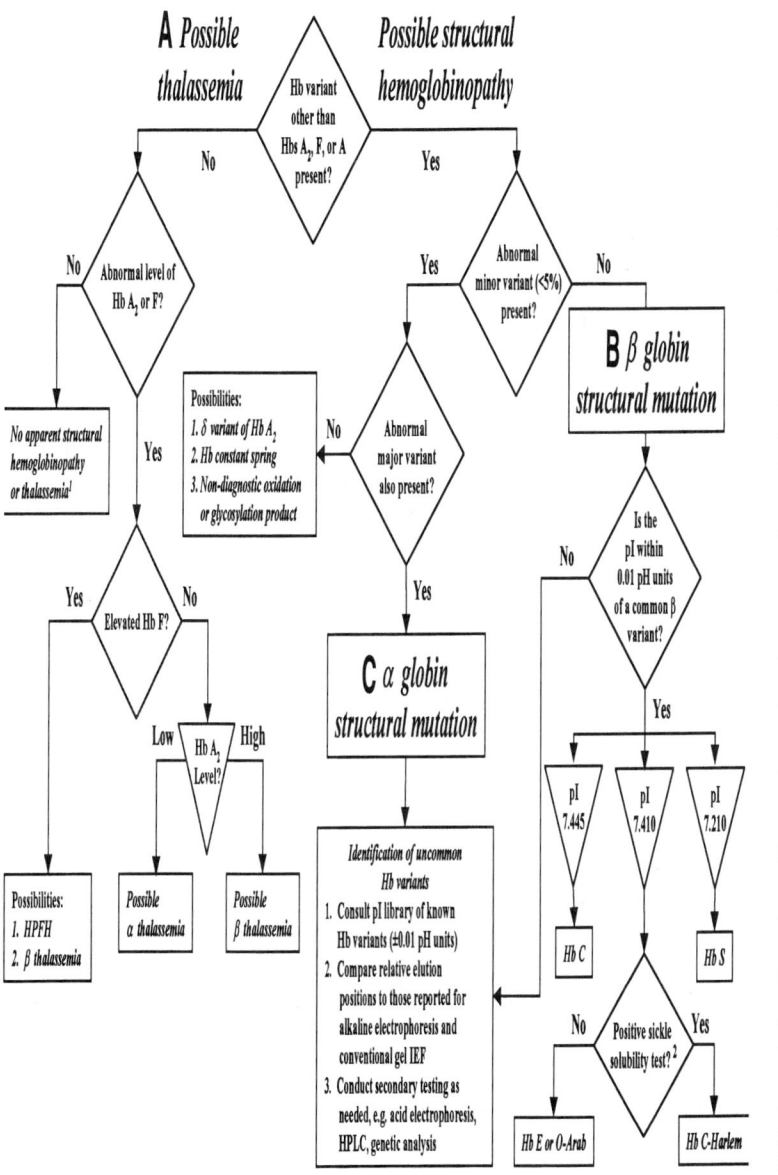

Fig. 4. cIEF interpretation flowchart. Three principal pathways are shown in the interpretation of cIEF results in the diagnosis of common congenital Hb disorders: **(A)** possible thalassemia; **(B)** β-globin structural mutation; **(C)** α-globin structural mutation. Interpretations can differ markedly for infants and compound heterozygotes.

[1]Heterozygous δβ-thalassemia and α-thalassemia are not excluded from the differential diagnosis.
[2]Only in the absence of HbS.

3. Similarly, the amounts of HbF and -F_{ac} (acetylated HbF, also called HbF_1, and usually present at ~10% of HbF) should also be combined in the absence of HbA. When HbA is also present, e.g., in neonates, HbF_ac partially co-migrates with HbA_1c. In this case, the percentage of HbA_1c/F_ac can be added to total HbA without affecting the interpretation.

4. As with conventional HbA_2 assay methods, HbA_2 cannot be precisely measured in the presence of HbC, because the two proteins are not completely resolved by cIEF (**Fig. 3**). Thus, HbA_2 should not be reported in patients with HbC, and any partially resolved HbA_2 should be added to the measured amount of HbC.

5. Although HbA_2 has a pI similar to those of HbE, -C-Harlem, or -O-Arab, it should not be confused with these major Hb variants. This is because HbA_2 is present at <10% of total Hb, even in individuals with β-thalassemia. In contrast, each of the above β-variants are usually present at levels well exceeding 20%, even in individuals with α-thalassemia. These major variants may be present at lower than 10% in infants, but HbA_2 levels are developmentally low and variable at this age, and are not diagnostically relevant. Elevated HbA_2, with HbF low or absent, and with evidence of anemia, microcytosis, and erythrocytosis, is relatively conclusive evidence of β-thalassemia. Low HbA_2 may be associated with α-thalassemias, but there is insufficient data correlating low HbA_2 levels with α-thalassemia to permit a definitive diagnosis based on this information.

6. If a major Hb variant with a pI of ~7.410 is encountered, it is probably HbE, -C-Harlem, or -O-Arab. Because HbC-Harlem is a sickling hemoglobin, it can be eliminated from the differential diagnosis (in the absence of Hb S) by a negative sickle solubility test. Although the identity of the unknown Hb variant may be presumed based on the patient's ancestry and the geographic origin of these mutations, confirmatory testing by citrate agar electrophoresis is recommended to verify identification.

3.3.3. Identifying and Reporting Uncommon Hb Variants

The extent of the reportable information obtained by cIEF (and thus the need for confirmatory testing) depends in part on the experience of the laboratory performing the assay. Because cIEF for Hb variants is relatively new, there is limited information about the pI of many unusual Hb variants. Without this information, the identity of unknown Hb variants can only be inferred by comparing their relative focusing positions on cIEF to those reported by other methods. This may reduce the number of possible candidates, however, to the point a careful correlation of laboratory and clinical data with published reports can yield a good tentative diagnosis. The decision to accept this tentative diagnosis, or to pursue additional laboratory testing, must be made on a case-by-case basis.

The authors have reported the pI of a number of common and uncommon Hb variants analyzed by cIEF in this laboratory (*6*), and continue to add to this pI library using commercial controls, samples verified by other laboratories, or proficiency testing samples. The authors have also participated successfully in the College of American Pathologists Hemoglobinopathy Survey for several years, and have correctly identified many unusual variants, including Hbs Constant Spring, Koln, Zurich, Malmo, Lepore, and M-Saskatoon, among others, using the steps described below.

1. Perform a verification cIEF analysis for all samples containing uncommon Hb variants. In this analysis, program the instrument to inject (5 s) the abnormal control into the capillary behind the unknown sample hemolysate. After the assay is completed, enter the exact elution times for the known Hb variants in the control into the System Gold peak identification and mol wt tables, and recalculate the pI. In this way, the known Hb variants in the control serve as internal standards for very reproducibly calculating the pI of unknown peaks.

2. Compare the pI calculated in this manner to the pIs of known Hb variants determined previously by cIEF. In many cases, identification can be made with high probability if the pI of the unknown peak is within ~0.01 pH units of a previously identified Hb variant. For example, the authors have had sufficient experience with HbD-Los Angeles, HbG-Philadelphia, the common δ-variant of HbA_2 (HbA_2'), HbH, and HbBart's to report identification of these variants on the basis of the cIEF results alone.

3. If a match is not found in the pI library, compare the pI or relative elution position of the unknown Hb variant to those reported in the literature. The authors previously reported the pI of a number of abnormal variants analyzed by the cIEF method described in this chapter *(6)*. Relative focusing positions on cIEF can also be compared to those reported for gel IEF *(8–10)*, or relative elution positions observed on alkaline electrophoresis *(3)*.

4. The search can be further narrowed by determining if the unknown Hb variant is due to an α- or β-globin gene mutation. As indicated in **Fig. 4,** major α-variants are usually accompanied by a minor α-variant of HbA_2 (*see* **Note 9**). Also, most major α-globin variants comprise about 25% of total Hb (in the absence of concurrent α-thalassemia), reflecting the presence of the mutation in only one of the four α-globin genes *(2)*.

5. Samples containing unknown Hb variants can also be mixed and analyzed simultaneously with controls containing a known Hb variant with a similar pI. Evidence of two peaks in the region can eliminate the control Hb variant from the differential diagnosis. The presence of a single peak suggests that the unknown Hb and the control Hb variant are the same, and increases the confidence of the identification.

3.3.4. Summary

Capillary isoelectric focusing is a rapid primary assay for the sensitive detection and specific identification of many congenital Hb disorders. High resolution of abnormal major Hb variants makes the assay particularly useful for the diagnosis of common disorders like sickle cell disease and HbC disease, or for monitoring therapy for these diseases (e.g., measuring the decrease in HbS after hypertranfusion, or the increase in HbF following hydroxyurea chemotherapy). Precise quantification of minor variants makes the same assay useful for the diagnosis of thalassemia mutations. The need for confirmatory testing with other methods is limited, because cIEF can also identify many unusual Hb variants based on pI. More widespread use of the assay will increase the information available about the pI of many uncommon Hb variants, and enhance the specificity of the assay for the diagnosis of unusual structural hemoglobinopathies.

4. Notes

1. Conditioned with methanol before initial use, before each daily run, and between samples, a single DB-1 capillary (~$10, if purchased in bulk) will usually give reproducible results for several hundred samples. Some variability in performance between capillaries is encountered, however, possibly as a result of inconsistent internal coating efficiency during manufacturing. The authors have had capillaries last over 6 mo when used 3× a week, but also one capillary that proved unsuitable for even a single analysis. Capillary suitability or failure is best detected by a loss of quantification accuracy when evaluating the controls. An early indicator of capillary failure is an increase in the difference between the elution times (Δt, normally, about 4.4 min) of HbA_1c and HbA_2 (*6*). The authors agree with others that the DB-1 is not an ideal capillary, but it is unclear that a more cost-effective capillary with better long-term performance exists.

2. The authors recommend preparation of larger volumes of ampholyte–methylcellulose solution for extended use, rather than daily preparation of small volumes, because small differences in ampholyte concentration (e.g., 2.0 vs 2.5% v/v) markedly affect the distribution of focused proteins inside the capillary. Small pipeting or measuring errors are magnified in the preparation of small volumes, and can markedly affect elution time and day-to-day reproducibility, especially with viscous solutions, such as the commercial ampholytes and the methylcellulose solution. Furthermore, the ampholyte–methylcellulose solution must be prepared in advance, because the distribution of focused proteins is different (wider) in freshly prepared solution, compared to that prepared at least 1 d prior to use. Under normal use (0.2 mL/d), a 5-mL batch of ampholyte–methylcellulose solution will last for about 25 working days.

3. The authors recommend caution if using a diode-array detector to perform cIEF with the reagent and capillary system described here. The results of some early experiments in this laboratory showed that only two or three assays could be performed with DB-1 capillaries installed on a P/ACE instrument equipped with a diode-array detector. The same capillaries could not subsequently be used successfully on an UV system. The reason for the capillary failure remains unclear.

4. For the Beckman P/ACE, water-filled vials should be kept in tray positions 1 and 11, because these are the default capillary docking positions in the event of a

system power failure. It is therefore practical to also use these vials as the docking sites for the capillary when the instrument is not in use.

5. Hemolysates prepared from RBC give more reproducible results than those prepared from whole blood *(6)*. Depending on the need, however, suitable results may be obtained using whole blood or blood collected on filter paper. Washing the RBC in saline, or centrifuging the hemolysate before use, is not necessary.

6. Optimal sample injection time can vary, depending on a number of factors. Most notably, transmittance may be slightly different with different capillaries, probably because of variations in the internal diameter, or how neatly the window was prepared and aligned in the cartridge. The authors change the sample injection time a few seconds (infrequently necessary) to obtain peak heights in controls similar to that previously encountered with other capillaries.

7. All of the common normal and abnormal Hb variants have pIs higher than that of HbA_1c. Some uncommon abnormal hemoglobins have much lower pIs, however, including HbsH and -Bart's. Extending the elution time from 10 to 20 min allows detection of all known abnormal Hb variants. Unfortunately, this would markedly lengthen the assay time for most samples that do not contain unusual low pI Hb variants. The authors have observed that Hb variants that have not eluted from the capillary during the normal run time show up as small peaks in the methanol rinse. Careful inspection of the detector output at the end of the run, therefore, can limit the need for longer focusing times. Certainly, any Hb variant comprising over 10% of total Hb can be detected in this manner. Because HbBart's is most common at low levels in neonates, it is the authors' policy to analyze all samples from patients under 30 d of age, using the longer elution time.

8. It should be remembered that System Gold peak integration software is not specifically designed for this particular use. As a consequence, autointegration sensitivity must be adjusted to omit the detection of very small peaks attributable to minor posttranslational Hb products that do not contribute to the diagnosis of congenital Hb disorders. Also, one set of integration parameters does not necessarily work best for all sample conditions. It is therefore sometimes expedient and clinically acceptable to manually adjust peak integration. The authors specifically avoid manually adjusting peak integration for HbA_2, however, to avoid bias in the evaluation of thalassemia syndromes.

9. In some cases, the associated α-variant of HbA_2 may co-elute with another Hb variant, and not be detected (as appears to be the case with HbI). Care should also be exercised not to confuse α- and δ-variants of HbA_2. A common δ-globin variant of HbA_2 is observed in about 2% of African-Americans *(1)*.

References

1. Fairbanks, V. F. (1980) *Hemoglobinopathies and Thalassemias,* Brian C. Decker, New York.
2. Bunn, H. F. and Forget, B. G. (1986) *Hemoglobin: Molecular, Genetic and Clinical Aspects.* W. B. Saunders, Philadelphia, PA.

3. Schmidt, R. M. (1993) Laboratory diagnosis of hemoglobinopathies, in *Hematology: Clinical and Laboratory Practice* (Bick, R. L., ed.), CV Mosby, St. Louis, MO, pp. 327–390.

4. Hempe, J. M. and Craver, RD. (1994) Quantification of hemoglobin variants by capillary isoelectric focusing. *Clin. Chem.* **40,** 2288–2295.

5. Craver, R. D., Abermanis, J. G., Warrier, R. P., Ode, D. L., and Hempe, J. M. (1997) Hemoglobin A_2 levels in healthy persons, sickle cell disease, sickle cell trait, and β thalassemia by capillary isoelectric focusing. *Am. J. Clin. Pathol.* **107,** 88–91.

6. Hempe, J. M., Granger, J. G., and Craver, R. D. (1997) Capillary isoelectric focusing of hemoglobin variants in the pediatric clinical laboratory. *Electrophoresis* **18,** 1785–1795.

7. Hempe, J. M., Granger, J. G., Warrier, R. P., and Craver, RD. (1997) Analysis of hemoglobin variants by capillary isoelectric focusing. *J. Capillary Electrophoresis* **4,** 131–135.

8. Basset, P., Beuzard, Y., Garel, M. C., and Rosa, J. (1978) Isoelectric focusing of human hemoglobin: Its application to screening, to the characterization of 70 variants, and to the study of modified fractions of normal hemoglobins. *Blood* **51,** 971–982.

9. Koepke, J. A., Thoma, J. F., and Schmidt, R. M. (1975) Identification of human hemoglobins by use of isoelectric focusing in gel. *Clin. Chem.* **21,** 1953–1955.

10. Black, J. (1984) Isoelectric focusing in agarose gel for detection and identification of hemoglobin variants. *Hemoglobin* **8,** 117–127.

11

Serum Apolipoproteins

Layle K. Watkins, Steven L. Cockrill, and Ronald D. Macfarlane

1. Introduction

The apolipoproteins associated with serum lipoprotein particles give structural stability as well as regulatory control in lipid metabolism. The development of atherosclerosis is linked to dysfunction in lipid metabolism, and the serum lipoproteins are directly involved, either through the action of their apolipoprotein components or in combination with the lipids sequestered in these particles. Consequently, quantitative features of the apolipoproteins (apos), serum levels, distribution within the lipoprotein population, and the appearance of isoforms, are potential markers for cardiovascular disease risk that are being added to the lipid profile for more effective cardiovascular risk screening (1). Currently, apo quantitation is being carried out using immunoassay or gel electrophoresis for separation and staining with image analysis for quantitation (2). Capillary electrophoresis has the potential for higher resolution, greater specificity, speed, and automation, coupled with on-line detection for more accurate and precise quantitation.

Each of the three major classes of lipoproteins, very-low density lipoprotein (VLDL), low-density lipoprotein (LDL), and high density lipoprotein (HDL), contain major apos that are characteristic of each class. VLDL and LDL particles contain a single apolipoprotein molecule (apo B-100; mol wt 512,932), and each HDL particle contains at least one apo molecule (apo A-I, mol wt 28,079). In addition, other apos are found in varying amounts, and include the C apos and apo E found in VLDL and HDL, and apo A-II in HDL. The goal of the development of the CE-based protocol in this laboratory is to quantitate the levels of these apos in the three lipoprotein fractions, as a component of developing a comprehensive cardiovascular risk profile (3–6).

From: *Methods in Molecular Medicine, Vol 27: Clinical Applications of Capillary Electrophoresis*
Edited by: S. M. Palfrey © Humana Press Inc., Totowa, NJ

In the authors' method, adsorption of the apos on the silica capillary wall is minimized by operating at a basic pH, of which the walls of the silica capillary are negatively charged because of silanol deprotonation, and the apos pI values *(5–6)*, being negatively charged, are repelled from the capillary wall during electromigration. Separation of the apos under CE conditions is based on affinity for a mixture of dodecyl and tetradecyl sulfate anions modulated by an organic modifier, acetonitrile *(3)*. In this environment, each of the apo species form an anion complex with a characteristic charge/volume value, and can be identified and quantitated in the electropherogram, based on measured effective electrophoretic mobility and UV-absorbance.

The first step in the analysis is separation of the lipoprotein classes by fast ultracentrifugation, using a self-generating sucrose density gradient *(6)*. The lipoprotein layers are recovered from the ultracentrifuge tube, and the total protein of each lipoprotein fraction is measured *(6)*, then analyzed by CE, using two different delipidation protocols to separate the apo components from the lipids. The first method uses the CE buffer for delipidation of the LP fraction, and the CE analysis is carried out using the mixture of delipidated apos and lipids. For VLDL and LDL, apo B-l00 is identified in the electropherogram and the free lipids complex with SDS, to produce a distribution of anions of triglycerides for VLDL; the cholesterol/cholesterol ester components of LDL form a negatively charged complexes with a characteristic electrophoretic mobility *(6)*. The action of the CE buffer on HDL produces species that have been identified in the electropherogram as free apo A-I, II and apo A-I, II associated with residual lipid. The second delipidation method completely separates the apos from lipids, and uses solid-phase extraction (SPE). The apo B-l00 component is retained in the SPE cartridge, and the remainder of the apos are recovered and analyzed by CE. High-resolution electropherograms are obtained that are used to identify the apos present, and to measure how the total protein content is distributed within the apo population *(5)*.

Included in this chapter is an example of the application of the authors' CE-based methodology to an ongoing clinical study of short-term changes in the lipoprotein profiles of acute-phase myocardial infarction (MI) patients. The patient selected presented with an inferior MI. Blood samples remaining after measurement of cardiac enzyme levels in the hospital's pathology laboratory were also analyzed by the CE-based protocol. Samples were drawn over a 3-d period, with the first sample obtained within a few hours of the MI event. During that period, the patient was treated with heparin and tissue plasminogen activator, and a cardiac catheterization was performed with right coronary artery stent placement. The objective of the apo analysis was to determine whether any changes in the profile could be detected during the 3-d period. Significant changes were observed in this initial study, warranting further

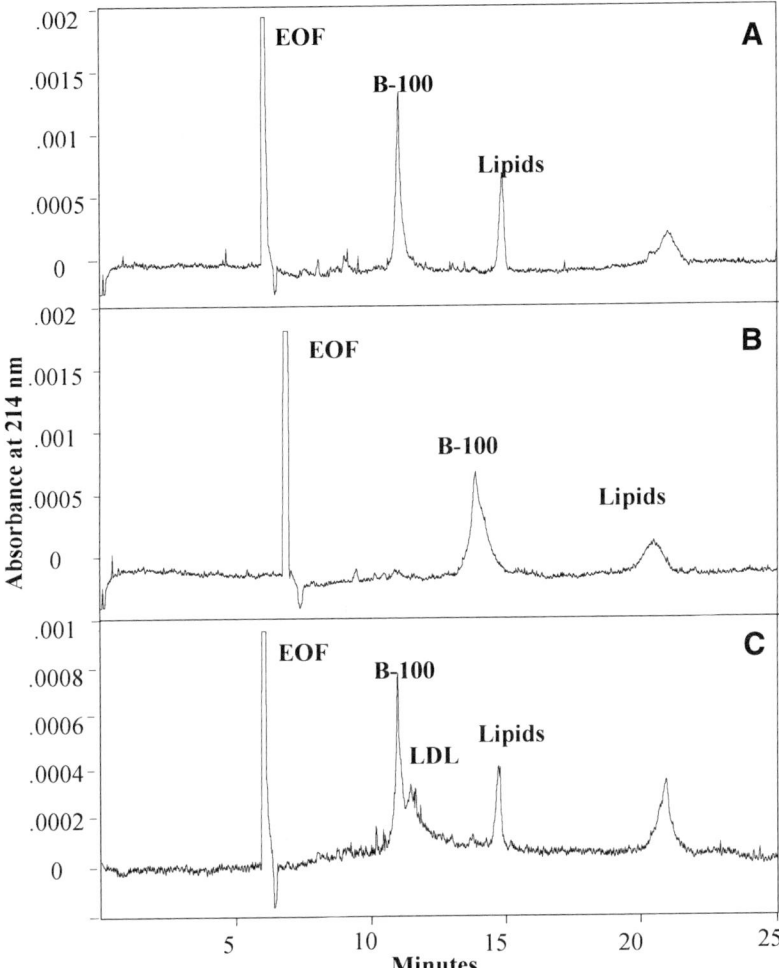

Fig. 1. Electropherograms of the LDL fraction from the serum of a post-MI patient. **(A)** Sample obtained in first 24 h of hospitalization; **(B)** d 2 sample; **(C)** d 3 sample.

development of the study to understand the origin of these changes, and to determine relevance to the pathology and treatment. Three sets of electrophero-grams were obtained, and are shown in **Figs. 1–3**. Further details and comments are given in the figure captions. The purpose of including these preliminary findings in this chapter is to demonstrate how an established medical protocol, carried out by medical staff at a hospital, can be supported in a parallel protocol at the molecular medicine level, using modern methods of analysis that include CE.

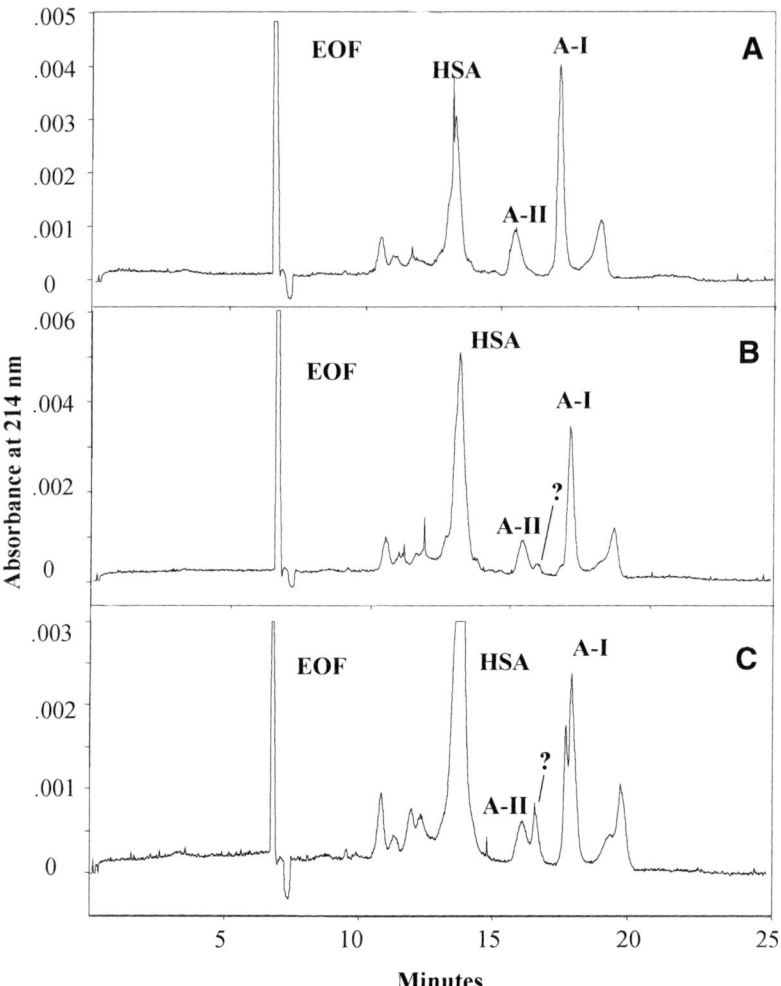

Fig. 2. Electropherograms of the HDL fraction from the same patient and time sequence as for **Fig. 1,** using the CE buffer to delipidate the sample. (**A**) d 1; (**B**) d 2; (**C**) d 3.

2. Materials

1. Density gradient medium: The medium is prepared in bulk to minimize variability. A weight of 20.0 g sucrose is dissolved in 80.0 g Milli Q deionized water (ddH$_2$O; Millipore, Bedford, MA) to produce a 20% w/w solution. To preserve the shelf life, solid sodium azide is added to the solution to yield a 0.1% concentration. The medium solution is stored at room temperature indefinitely (*see* **Note 1**).

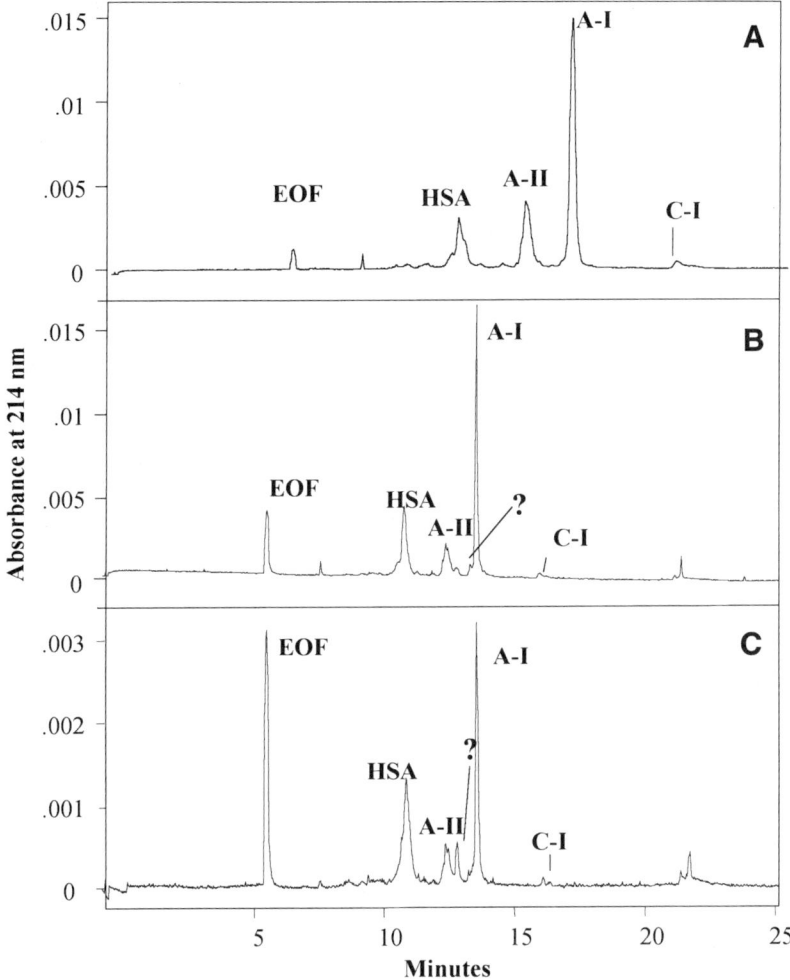

Fig. 3. Electropherograms of the HDL fraction from the same patient and time sequence as for **Figs. 1** and **2,** but using the SPE cartridge for delipidation. (**A**) d 1; (**B**) d 2; (**C**) d 3.

2. Lipophilic stain: The lipoproteins are identified in the ultracentrifuge (UC) tube by the incorporation of a lipid soluble stain; Sudan Black B (SBB). The stain solution is prepared by dissolving 0.1 g SBB in 10.0 mL dimethyl sulfoxide (DMSO). Complete dissolution is aided by sonication, and the solution is stored at room temperature.

3. Protein assay *(7)*: Bicinchoninic acid (BCA) standard working reagent (SWR): The SWR is prepared by mixing 50 volumes of reagent A (1% BCA,

2% $Na_2CO_3 \cdot H_2O$, 0.16% sodium tartrate, 0.4% NaOH and 0.95% $NaHCO_3$) (Pierce, Rockford, IL) with 1 vol of reagent B (4% $CuSO_4 \cdot 5H_2O$). Reagents A and B can be stored at room temperature indefinitely. For best results, fresh SWR should be prepared daily (*see* **Note 2**).

4. Bovine serum albumin (BSA) standards: A set of BSA standards (0–2 mg/mL) is prepared with ddH_2O. These standards generate the calibration curve for the BCA protein assay.

5. CE electrolyte: The background electrolyte (BGE) is prepared fresh daily by mixing 50 m*M* sodium borate, 3.5 m*M* 70% sodium dodecyl sulfate (SDS; Sigma, St. Louis, MO), and 20 mL acetonitrile (ACN), and diluting to 100 mL with ddH_2O. The electrolyte is filtered using a 0.22-μm syringe filter. Degassing by vacuum and sonication is carried out to avoid bubble formation during the run (*see* **Note 3**).

6. SPE mobile phase: The SP mobile phases, HPLC grade H_2O (mobile phase A) and ACN (mobile phase B), are acidified with 0.1% v/v trifluoroacetic acid (TFA).

7. SPE cartridges: Sep Pak tC18 light cartridges are purchased from Waters (Milford, MA).

8. CE capillary: An untreated fused silica capillary (Polymicro Technologies, Phoenix, AZ), 75 μm id, 375 μm od, is used. The total length of the capillary is 56.6 cm, with an effective length of 50.4 cm.

3. Method

3.1. Sample Preparation

1. Draw blood samples into a Vacutainer brand blood collection tube, with no additives. Allow the blood to clot at room temperature, before centrifugation at 1000*g* for 20 min. Withdraw the serum from the top, above the pelleted clot and red blood cells, and store at 4°C.

2. Incubate a 400-μL aliquot of serum with 5 μL SBB (1% w/v in DMSO) for 30 min at room temperature. Briefly mix the sample to ensure homogeneity. Pellet any excess stain at the bottom of the sample by centrifugation at 1000*g* for 15 min.

3. Layer a 400-μL vol of the stained serum sample above 800 μL 20% sucrose in a 1.5-mL capacity, thick-walled, open-topped polycarbonate ultracentrifuge tube. Separate the lipoproteins into bands by centrifugation using the Optima TLX tabletop ultracentrifuge equipped with a TLA 120.2 rotor (Beckman, Fullerton, CA) at 627,000*g* and 20°C for 250 min. Set the acceleration and deceleration rates at 5 and 0 min (coasting stop without brake), respectively.

4. Upon completion of the separation, the lipoprotein fractions (VLDL, LDL, and HDL) are visible as blue bands in the transparent medium. Withdraw them from the density gradient by aspiration with a pipet.

5. Mix a 10 μL vol of each UC fraction and BSA standard with 1 mL BCA-SWR (*see* **Note 4**). Vortex the samples/standards, and incubate them at 60° C for 30 min, and then cool to room temperature (5 min).

6. Measure the absorbance of each sample/standard at 562 nm using a Perkin-Elmer Lambda 4B UV/Vis spectrophotometer (Norwalk, CT), autozeroed using ddH_2O. Subtract the background absorbance using the 0 mg/mL standard (ddH_2O in

SWR), and determine the concentration of each fraction using the BSA standard calibration curve (*see* **Note 5**).

7. Partial delipidation of the UC fraction by SDS is accomplished by mixing 2 μL of each fraction (LDL and HDL) with 20 μL of the BGE, and incubating at 37° C for 30 min (*see* **Note 6**). Cool the SDS delipidated samples to room temperature (5 min), and then degass by centrifugation at 7500*g* for 5 min.

8. Prepare the HDL fraction for SPE delipidation by acidifying an aliquot (~200 μg protein) with 0.1% (v/v) TFA.

9. Condition the SPE cartridge by rinsing with 5 mL mobile phase B, followed by 5 mL mobile phase A applied by syringe at a flow rate of 10 mL/min. Take care to avoid the introduction of air into the cartridge prior to the sample loading (*see* **Note 7**).

10. Load the acidified sample (**step 8**) by pipet into a syringe barrel, and apply to the SPE cartridge by syringe at a flow rate of 1 mL/min.

11. Rinse the SPE cartridge with mobile phase A at a flow rate of 1 mL/min, and then purge with air (*see* **Note 8**). Elute the proteins using a series of 50-μL rinses of mobile phase B applied by syringe. Each time, purge the cartridge with air between eluent applications. Collect the third aliquot, and perform the BCA test as described in **steps 5** and **6**, to determine the protein concentration (*see* **Notes 9** and **10**). Add a 10 μL vol of rinse no. 3 to 10 μL ddH$_2$O, mix, and centrifuge as describe in **step 7** (*see* **Note 11**).

3.2. Capillary Electrophoresis

1. Separate the SDS and the SPE delipidated samples using a Beckman P/ACE model 5510 CE instrument (Beckman) equipped with a diode array detector. P/ACE Station Migration software (Beckman) is used to control the system, and for analysis. The separation is carried out at a constant voltage of 17.5 kV, current of 24 μA, and at 20°C. The electropherograms are monitored at 214 nm. The sample is injected using a 4-s pressure injection (0.5 psi), and the electroosmotic flow (EOF) is monitored for each analysis, using a separate 1-s injection of ddH$_2$O following the sample injection (*see* **Note 12**).

2. LDL and HDL electropherograms, using SDS delipidation, are shown in **Figs. 1** and **2**, respectively. An electropherogram of HDL delipidated by SPE is shown in **Fig. 3** (*see* **Note 13**).

3. **Figure 1** shows electropherograms of the LDL fraction from the serum of a post-MI patient. Delipidation was carried out in the CE buffer. (**A**) Sample obtained in first 24 h of hospitalization. The peak labeled "EOF" is the EOF marker from H$_2$O; "B-100" is an SDS/apoB-100 anion complex, and "lipids" is a complex of lipids and SDS. The unlabeled peak on the far right is caused by the SBB (SBB) stain used as an ultracentrifugation marker. (**B**) Day 2 sample; note that the EOF marker has shifted to a longer time (*see* **Note 13**). The weaker EOF results in a significant time shift in the "B-100," and "lipids" peak and the SBB peak is shifted out of the 25-min time window. (**C**) Day 3 sample; the "B-100" and "lipids" and SBB peaks are seen within the 25-min window. For reasons not known, the LDL

was not fully delipidated in this run, even though delipidating conditions were the same as the other two runs. The residual LDL peak is labeled "LDL."

4. This set of runs shows the influence of capillary degradation on the electropherograms. The position of the EOF marker is the diagnostic for capillary performance. Even though a varying EOF results in shifts in the mobilities of the peaks, the effective mobilities remain the same.

5. **Figure 2** shows electropherograms of the HDL fraction from the same patient and time sequence as for **Fig. 1**, using the CE buffer to delipidate the sample. (**A**) The dominant peaks are caused by human serum albumin contamination (HSA), apo A-Il, and apo A-I. The peaks between HSA and the EOF marker are from other serum proteins. The peak on the high time side of apo A-I is from partially delipidated apo A-I *(4)*. (**B**) D 2, the electropherogram is similar to that for d 1, but a new peak is present between apo A-Il and apo A-I. (**C**) D 3 shows a considerable increase in the concentration of this new component, as well as the conversion of the apo A-I peak to a doublet. Changes in the particle density profile and mass spectra were also observed. The mass spectrum identifies this protein as one of the serum amyloid proteins A(5AA), (apoJ) an acute phase response protein.

6. **Figure 3** shows electropherograms of the HDL fraction from the same patient and time sequence as for **Figs. 1** and **2**, but using the SPE cartridge for delipidation. (**A**) D 1, The HSA contamination level is considerably reduced, and the apo A-I, II peaks are sharper, with the elimination of the lipids from the sample. With the improved background, a minor component, apo C-I, has been identified, based on its characteristic effective electrophoretic mobility. (**B**) D 2, For this run, a new capillary was used. Note the shift in EOF and all other peaks to shorter times, and with improved resolution. The apoJ peak between apo A-Il and apo A-I, observed in the HDL/SDS delipidation study (**Fig. 2**), is evident in this electropherogram. (**C**) D 3, The electropherogram shows the apoJ peak at a higher intensity and well resolved. The temporal changes in this sequence of electropherograms is linked to the physiological response of the patient to the MI event, and, although this study is in the beginning stages, a sense of closeness between clinical manifestations and molecular medicine is emerging.

4. Notes

1. All solutions involved in the ultracentrifugation sample preparation must be thoroughly mixed prior to use.
2. One study states that the SWR is usable for 1 wk when stored at room temperature. However, a drastic decrease in sensitivity was observed after only a few hours at room temperature. When the SWR was stored at 4°C, it is usable for 2 d, with only a slight decrease in sensitivity. The best sensitivity is observed when the SWR is prepared just prior to use.
3. When using the BGE, to reduce/eliminate the number of sharp peaks appearing in the electropherogram because of the formation of gas bubbles, it is essential to prepare, filter, and degas the BGE each day. For best results, the BGE vials should be replaced with freshly degassed electrolyte every 4–6 h.

4. The manufacturer recommends mixing 50 μL of each sample/standard with 1 mL SWR for the BCA assay. Because of sample limitations, 10 μL of each sample/standard is used, producing a calibration curve with a dynamic range of 0.15–4.00 mg/mL, when the spectrophotometer is zeroed using ddH$_2$O.

5. The least amount of variation in the BCA calibration curve (slope and intercept) was observed when the spectrophotometer was zeroed using ddH$_2$O, and the standards were background subtracted using a 0 mg/mL (ddH$_2$O in SWR) standard.

6. A 30-min incubation period is sufficient for delipidation when the lipoprotein fractions are first removed from the UC. Longer incubation periods (2–4 h) are necessary when the lipoprotein fractions are stored at 4° C overnight.

7. It is important that the SPE cartridge is not exposed to air during the conditioning steps. Partial drying of the stationary phase can drastically affect the loading capacity and analyte recovery.

8. After the acidified HDL sample is loaded and any hydrophilic impurities are eluted with the 0.1% aqueous TFA rinse, the SPE cartridge is purged with air to remove as much mobile phase A as possible. This step influences the reproducibility of the protein elution.

9. Aliquot no. 3 is collected and analyzed, because the majority (50–70%) of the eluted proteins are in this aliquot. If the HDL is contaminated with albumin (HSA), the majority of the HSA is eluted in aliquot no. 2, markedly decreasing the interference by HSA in the HDL electropherogram.

10. The BCA test is repeated so that the amount of protein in aliquot no. 3 is known, and can be either concentrated or diluted, as necessary. It was determined that 0.5–1 mg/mL is an ideal range for the apo detection by CE, using this BGE.

11. It is necessary to dilute the ACN–protein solution (aliquot no. 3) with ddH$_2$O prior to injection, to maintain a stable current and eliminate current errors caused by the decreased conductivity of the sample plug.

12. An internal EOF marker is necessary to obtain reproducible effective mobilities. Benzyl alcohol, DMSO, and acetone were examined as internal EOF markers, but were found to be unacceptable.

13. The influence of the condition of the capillary on the quality of the electropherogram is demonstrated by comparing **Fig. 3A** with **Fig. 3B**. The capillary used to obtain the electropherogram shown in **Fig. 3B** was new; the capillary used to obtain the electropherogram shown in **Fig. 3A** had been used for numerous runs. Note the time shift of the EOF marker between the two runs, indicating the degradation of the capillary. With the new capillary, the EOF marker migrated at shorter times, indicating a stronger EOF, and the peak resolution was considerably improved. The electropherogram seen in **Fig. 1B** was run after **1C**, and just prior to **3A**, and the EOF is also shifted, because of the poorer capillary performance.

Acknowledgments

This work was supported by the U.S. National Institutes of Health, Heart, Lung and Blood Institute (HL 54566). The authors are grateful for the assistance of the medical staff at Scott and White Hospital, and, in particular, Dr. Eugene Terry and Dr. Catherine J. McNeal.

References

1. Mahley, R. W., Innerarity, T. L., Rail, S. C., Jr., and Weisgraber, K. H. (1984) Plasma lipoproteins: apoprotein structure and function. *J. Lipid Res.* **25,** 1277–1294.
2. Bergeron, N., Kotite, L., and Havel, R. J. (1996) Simultaneous quantification of apolipoproteins B-i 00, B-48, and E separated by SDS-PAGE. *Methods Enzymol.* **263,** 82–94.
3. Cruzado, I. D., Hu, A. Z., and Macfarlane, R. D. (1996) Influence of dodecyl sulfate ions on the electrophoretic mobilities of lipoprotein particles measured by HPCE. *J. Capillary Electrophoresis* **3,** 25–29.
4. Cruzado, I. D., Song, S., Crouse, S. F., and O'Brien, B. C. (1996) Characterization and quantitation of the apoproteins of high-density lipoprotein by capillary electrophoresis. *Anal. Biochem.* **243,** 100–109.
5. Macfarlane, R. D., Bondarenko, P. V., Cockrill, S. L., Cruzado, I. D., Koss, W., McNeal, C. J., Spiekernian, A. M., and Watkins, L. K. (1997) Development of a lipoprotein profile using capillary electrophoresis and mass spectrometry. *Electrophoresis* **18,** 1796–1806.
6. Cruzado, I. D., Cockrill, S. L., McNeal, C. J., and Macfarlane, R. D. (1998) Characterization and quantitation of apolipoprotein B-100 by capillary electrophoresis. *J. Lipid Res.* **39,** 205–217.
7. Smith, P. K., Krohn, R. I., Hermanson, G. T., Mallia, A. K., Garmer, F. H., Provenzano, M. D., et al. (1985) Measurement of protein using bicinchoninic acid. *Anal. Biochem.* **150,** 76–85.

12

Gene Dosage in Capillary Electrophoresis

Prenatal Diagnosis of Down's Syndrome
and Rh D/d Genotyping

Pier Giorgio Righetti, Cecilia Gelfi, and Gian Franco Cossu

1. Introduction

The first instrumentation for capillary zone electrophoresis (CZE) became available 10 years ago, and, since that time, interest in the technique has grown at a steady pace. However, although CZE has been embraced by a diverse scientific community that includes biochemists, chemists, and molecular biologists, acceptance of CZE by the clinical laboratory community has been markedly slower. Nevertheless, two recent monographs, devoted to this particular topic, testify to the growing interest of CZE in the clinical science *(1,2)*. In these two special issues, some reviews are devoted specifically to the topic of DNA analysis *(3–6)*. Another issue of *Electrophoresis,* dedicated to reviews on general topics of CZE, has recently appeared *(7)*. An additional number of recent reviews testify to the great interest in the use of CZE for analysis of DNA for diagnostic purposes *(8–13)*.

It must be emphasized that, in general, DNA analysis by CZE for molecular diagnostics has been adopted mostly as an analytical tool. The diagnostic value of the CZE profile is attributed to the disappearance of one (or more) peaks present in the normal control, and to the appearance of one (or more) new peaks in the disease conditions under study. The new peaks typically have a different mobility, caused either by changes in size of the DNA analyte (e.g., small deletions, modification of a cutting site) or by the presence of a single-base mutation (however, in this last case, particular electrophoretic approaches have to be adopted, such as single strand conformation polymorphism [SSCP] or denaturing gradient gel electrophoresis [DGGE]). Typical examples can be found, e.g., in cystic fibrosis *(14)* and in Duchenne/Becher muscular distrophy

From: *Methods in Molecular Medicine, Vol 27: Clinical Applications of Capillary Electrophoresis*
Edited by: S. M. Palfrey © Humana Press Inc., Totowa, NJ

(15). Only a few authors have addressed the problem of gene quantitation, i.e., quantitative analysis of the peak of DNA analyte. Examples will be found in the two topics of the present chapter: diagnosis of Down's syndrome *(16)* and Rh D/d genotyping *(17)*. Other examples include diagnosis of follicular lymphomas *(18)*, hepatitis C virus *(19)*, polio virus *(20)*, HIV-1 virus *(21–23)*, and basic fibroblast growth factor in human ovarian carcinomas *(24)*.

Below, a brief description of both conditions will be given.

1.1. Down's Syndrome

Free trisomy 21 accounts for about 95% of all cases of Down's syndrome, the leading cause of genetically inherited mental retardation. This birth defect is among the most common genetic diseases, with an incidence of about 1 in 700 births, with an exponential increase in pregnancies in women over 35 yr of age. Prior to birth, Down's syndrome can be detected by amniocentesis or chorionic villus sampling and karyotyping, an expensive, labor-intensive, and time-consuming process. Thus, a less-expensive screening test indicating the presence of chromosomal abnormalities would be highly desirable. In 1991, Lubin et al. *(25)* described a quantitative polymerase chain reaction (PCR) amplification process for detecting X-chromosome dosage differences in patients with sex chromosomal aneuploidies. By assuming that, within the exponential phase of PCR amplification, the amount of specific DNA produced is proportional to the quantity of initial target sequences, this procedure was modified by Mansfield *(26)* for detecting trisomies 21 and 18, and the triple-X syndrome. The assay is based in extracting DNA either from blood or from amniotic cells, and in amplifying a highly polymorphic (90% heterozygosity), chromosome-21-specific D21S11 marker.

1.2. Rh Blood Group Antigens

The Rh (Rhesus) blood-group antigens are carried by a series of at least three homologous, but distinct, membrane-associated proteins. Two of these proteins have immunologically distinguishable isoforms, designated C,c and E,e, but the principal protein, D (or dominant antigen), has no immunologically detectable isoform d. The rhesus gene locus, on chromosome 1p34-p36, consists of two adjacent homologous structural genes, designated Rh CcEe and Rh D. An early and safe method for determining the zygosity for the D gene (DD or Dd) could have applications, for instance, in the prediction of the D genotype of a father in couples in whom there is an Rh D-negative woman at risk of fetal alloimmunization. Fifty-six percent of Rh-positive whites are in fact heterozygous for the Rh D antigen. When the father is heterozygous Rh D-positive, there is a 50% chance that the fetus will be Rh D-negative, and thus will be unaffected. There remains the problem of a simple and reliable method for identifying Rh D-positive fetuses. In 1993, two reports independently appeared that addressed this problem. In one, Bennett et al. *(27)* proposed fetal Rh D typing by amplifying specific regions of DNA isolated from amniotic fluid. With specific primers, they could amplify a 136-base-pair (bp) region common to the Rh CcEe and Rh D genes (exon 7). With another set of primers, they amplified a 186-bp region specific to the 3'-untranslated sequence (exon 10) of the Rh D gene. In Rh D-negative cases, only the 136-bp fragment is amplified; in Rh D-positive fetuses, both the 136-bp and 186-bp products are expressed. In another approach, Chërif-Zahar et al. *(28)* proposed determination of the zygosity for the RhD gene by Southern blot analysis. Genomic DNA from donors was digested with the *Hin*dIII restriction enzyme, and hybridized with the exon 4. Gene dosage effects were estimated by densitometric evaluation of the relative intensity of 6.7/7.3-kbp fragments (exon 4 probe) corresponding to the D gene.

2. Materials

2.1. PCR

1. Primers for Down's: for amplifying the STR region D21S11, use the VS17T#3 and VS17T#4 primers, as described by Sherma and Litt *(29)*.
2. Primers for Rh D/d typing: Use the primers A1, A2, A3, and A4, as described by Bennett et al. *(27)*.
3. PCR reaction buffer: 10 mM Tris-HCl, pH 8.3.
4. Taq polymerase (Perkin-Elmer).
5. DNA thermal cycler (model 2400 from Perkin-Elmer).

2.2. Electrophoresis

1. TEMED (N,N,N',N'-tetramethylethylene diamine).
2. 40% ammonium persulphate (freshly prepared).
3. Electrophoresis buffer: typically 89 mM Tris, 89 mM borate, 2 mM EDTA, pH 8.3, (TBE).

3. Methods

3.1. DNA Isolation and PCR for Down's Syndrome

1. Extract DNA from 2 mL whole blood samples of human donors (controls and postnatal diagnosis), or from 2 mL amniotic fluid, using uncultured cells (prenatal diagnosis), by the standard phenol–chloroform extraction procedure, and adjust the DNA concentration to 5 ng/μL.
2. PCR mix: Make up a total volume of 25 μL to contain 5 ng genomic DNA, 0.5 μM of each primer, 50 mM KCl, 10 mM Tris-HCl buffer, pH 8.3, 1.5 mM MgCl$_2$, 0.01% (w/v) gelatin, 200 μM of each dNTP, and 0.5 U Taq polymerase (Perkin-Elmer/Cetus).

3. PCR protocol: Initial denaturation is at 95°C for 5 min, followed by an initial annealing temperature of 62°C for 1 min, decreased by 1°C/cycle for the first 10 cycles, and held constant at 52°C for 30 s for the remaining 20 cycles. For all cycles, denaturation should be for 1 min at 95°C, and extension for 1 min at 72°C.

3.2. DNA Preparation and PCR for Rh D/D Typing

1. Extract DNA from 2 mL whole blood samples of human donors by the standard phenol–chloroform extraction procedure, adjust DNA concentration to 2 ng/μL.
2. PCR mix: Make up a total volume of 25 μL to contain 2 ng genomic DNA, 40 pM of each of the four primers, 50 mM KCl, 10 mM Tris-HCl buffer, pH 8.3, 1.5 mM MgCl$_2$, 9% DMSO, 50 μM of each dNTP, and 1.0 U Taq polymerase (Perkin-Elmer/Cetus).
3. PCR protocol: Perform 25 cycles of denaturation at 92°C for 15 s, primer annealing temperature for 15 s at 51°C, and primer extension for 30 s at 72°C with a DNA thermal cycler (model 2400 from Perkin-Elmer).

3.3. Preparation of Short-Chain Uncrosslinked Polyacrylamide (30)

Linear polyacrylamide of reduced chain length is synthesized by using isopropanol as a chain transfer agent for controlling the molecular mass of the product.

1. Eleven % acrylamide is dissolved in water containing 3% isopropanol. Degas for 10 min with a water pump. Add 1 μL pure TEMED (N,N,N',N'-tetramethylethylene diamine) and 1 μL 40% ammonium persulphate (freshly prepared) per mL of gelling solution.
2. Polymerize at 70°C in a thermostatic bath for 2 h.
3. Precipitate the polymer from the viscous solution by adding pure ethanol to 70% (v/v) final concentration. Wash the precipitate twice with 70% (v/v) ethanol.
4. Let the powder of polyacrylamide desiccate. For use in CZE, add the powder to the desired final concentration to the appropriate CZE buffer.

3.4. Desalting PCR DNA Fragments Obtained by PCR (see Note 1)

3.4.1. Via Microcon/Micropure Centrifugal Systems

1. Pour the analyte (50 μL) into the reservoir of a Micropure, placed onto a Microcon 30 tube, which in turn sits on a Microcon vial (from Amicon, Beverly, MA).
2. Centrifuge at 8000g for 10 min. The concentrated, desalted sample remains in the upper Microcon. The filtrate that collects in the bottom Microcon vial contains excess solvent and salts, and should be discarded.
3. Remove the Micropure reservoir from the Microcon, add 450 μL water to the Microcon sample reservoir, and spin again at 8000g for 10 min. About 10–20 μL of desalted and concentrated sample will remain in the Microcon.
4. Invert the Microcon reservoir and insert it in a new vial (a small Eppendorf tube). Spin again at 8000g for 5 min. The clean DNA filtrate will be located at the bottom of the Eppendorf tube. For more details, *see* **ref. 31**.

3.4.2. Via Dialysis on a Floating Membrane

1. Take a dry VS membrane (0.025 μm average pore diameter from Millipore, Bedford, MA), and let it float gently on the surface of distilled water poured on a glass watch, or on a Petri dish.
2. Gently pipet a drop (20–40 μL) of sample on the center of it, taking care to avoid disturbing the membrane or putting pressure on it, since it will risk sinking.
3. After 20 min of dialysis, recover the sample by suction into the tip of a Gilson pipet. For more details, *see* **Fig. 1** in **ref. 32**.

3.5. Capillary Zone Electrophoresis

1. Use a fused-silica capillary with an inner coating for quenching the electroendo-osmotic flow (e.g., from Beckman [Fullerton, CA], or from Bio-Rad [Hercules, CA], both coated with polyacrylamide, or from Perkin-Elmer [Palo Alto, CA], the latter coated with polyvinyl alcohol [PVA]; *see* **Notes 2** and **3**).
2. Typical length is 37 cm (30 cm to the detection window) and id of the order of 75 or 100 μm (for higher detection sensitivity).
3. Fill it with running buffer (TBE) containing a viscous polymer solution, e.g., 6–8% polyacrylamide, prepared as in **Subheading 3.3.** (*see* **Notes 4–6**). In PVA-coated capillaries, TBE cannot be used, because it will form charged di-diol complexes; use a TAE buffer (Tris, acetate, EDTA of equivalent ionic strength; *see* **Note 7**).
4. Inject the desalted, concentrated PCR sample in the capillary (in polymer-filled capillaries only electrokinetic injection can be used). Typical conditions: 4000 V for 10–20 s (*see* **Note 8**).
5. In standard CZE units, equipped with UV/Vis detectors, and in the absence of fluorophores in the sample, set the detector wavelength at 260 nm (*see* **Notes 9** and **10**).
6. Running conditions: typically 100–200 V/cm, with transit times up to 50 min.
7. For accurate fragments sizing, the marker V DNA ladder (Boehringer Mannheim, Mannheim, Germany) can be sequentially injected or mixed with the sample (in the latter case, ascertain that there are no overlapping peaks).

3.6. Results

3.6.1 Analysis of Down's Syndrome

1. In the analysis of Down's syndrome, via polymorphic STR allelic markers, one expects the results to fall in to the following categories: (a) for normal individuals, the homozygote would exhibit a single, double intensity peak; the heterozygote would be resolved into two equal-intensity peaks (1:1 ratio). Trisomic individuals, on the contrary, are expected to fall into two major groups: those with three STR peaks of similar intensities (1:1:1 ratio; i.e., three different STR alleles), or those with two peaks with a ratio of 2:1 (i.e., two copies of an identical STR allele, and one of a different allele; however, in rare cases of homozygosity, a single peak might be obtained; in such cases, only karyotyping will give a correct diagnosis).

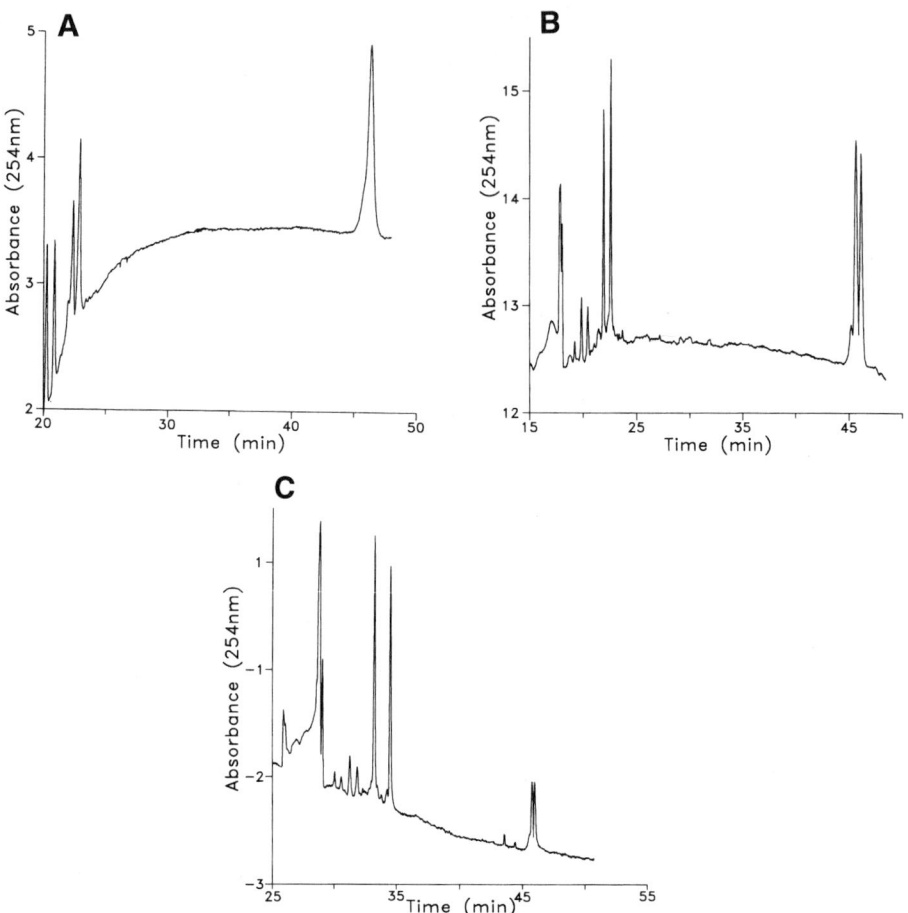

Fig. 1. CZE of PCR-amplified markers of Down's syndrome. Capillaries: 75 μm I.D. 37 cm length, coated with poly(N-acryolyl amino propanol) *(41)*. Background electrolyte: 89 mM Tris-borate, 2 mM EDTA, pH 8.3, containing 8% short-chain, low-viscosity, liquid sieving polyacrylamide and 2.5 μM ethidium bromide. Sample injection: 15 s, at 100 V/cm. Run: 100 V/cm, room temperature. **(A)** homozygous, healthy individual; **(B** and **C)** heterozygous, healthy individuals. Note, in **(B)** and **(C),** the two-peak pattern with a 1:1 gene ratio. All peaks eluting between 20 and 35 min represent primers and primer dimers. (Reproduced with permission from **ref. 6**.)

2. **Figure 1** shows the CZE analysis in sieving liquid polymers (8% short-chain, low-viscosity linear polyacrylamide) of some PCR amplified samples. Panel **(A)** shows the tracing of a homozygous, normal individual, exhibiting indeed a single peak. Panels **(B)** and **(C)** represent two heterozygous, normal individuals with two peaks of a ratio very close to unity. In all cases, the early eluting peaks (from

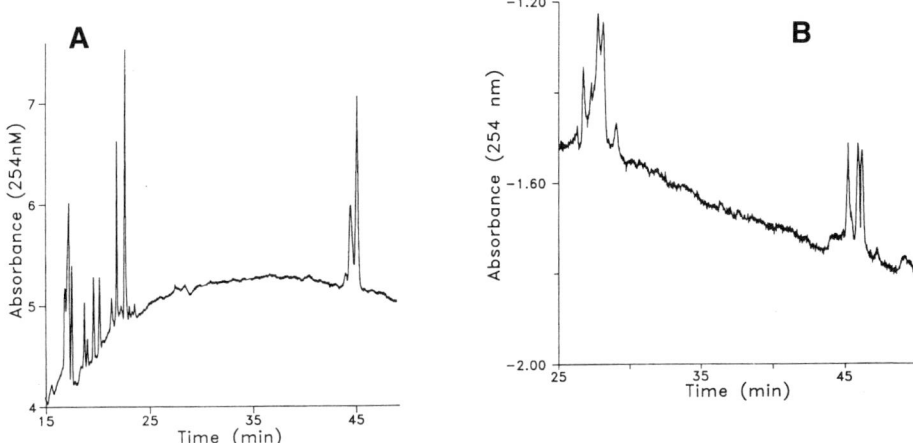

Fig. 2. CZE of PCR-amplified markers of Down's syndrome for prenatal diagnosis. All conditions as in Fig. 1, except that the analyte is a prenatal sample, with DNA isolated from amniotic cells. (**A**) Note the characteristic two-peak profile with a 2:1 gene expression ratio. (**B**) Note, in this particular sample, the characteristic three-peak profile with a 1:1:1 gene expression ratio (45–47 min migration window). (Reproduced with permission from **ref.** *6*.)

    20 to 35 min) represent excess primers (and primer-dimers); the diagnostic fragments have elution times in the 45–47 min window.

3. **Figure 2A** shows a prenatal diagnosis on DNA from amniotic fluid: The fetus, because of the characteristic 2:1 gene dosage, was diagnosed as affected by Down's syndrome, as independently confirmed by cytogenetic analysis, which indicated a case of trisomy 21. **Figure 2B** shows another case of prenatal diagnosis, in which the affected fetus exhibited a three-peak profile, with the characteristic 1:1:1 ratio.

3.6.2. Rh D/d Genotyping

1. According to the protocol of Bennett et al. *(27)*, Rh D is determined by amplifying a 136-bp region common to the Rh CcEe and Rh D genes, and a 186-bp region specific of the Rh D gene. Rh D positive and negative samples are then identified by polyacrylamide gel slab electrophoresis, followed by silver staining of the DNA bands.

2. CZE in sieving liquid polymers allows direct on-line peak densitometric evaluation by exploiting the intrinsic UV absorbance of the DNA fragments at 260-nm. The CZE method here demonstrated permits not only a rapid and reliable assessment of the Rh D type (the presence of both 136-bp and 186-bp fragments, indicating an Rh D-positive type; the presence of only the 136-bp fragment, indicating an Rh D-negative type, *see* **Fig. 3A**), but also a rapid determination of the zygos-

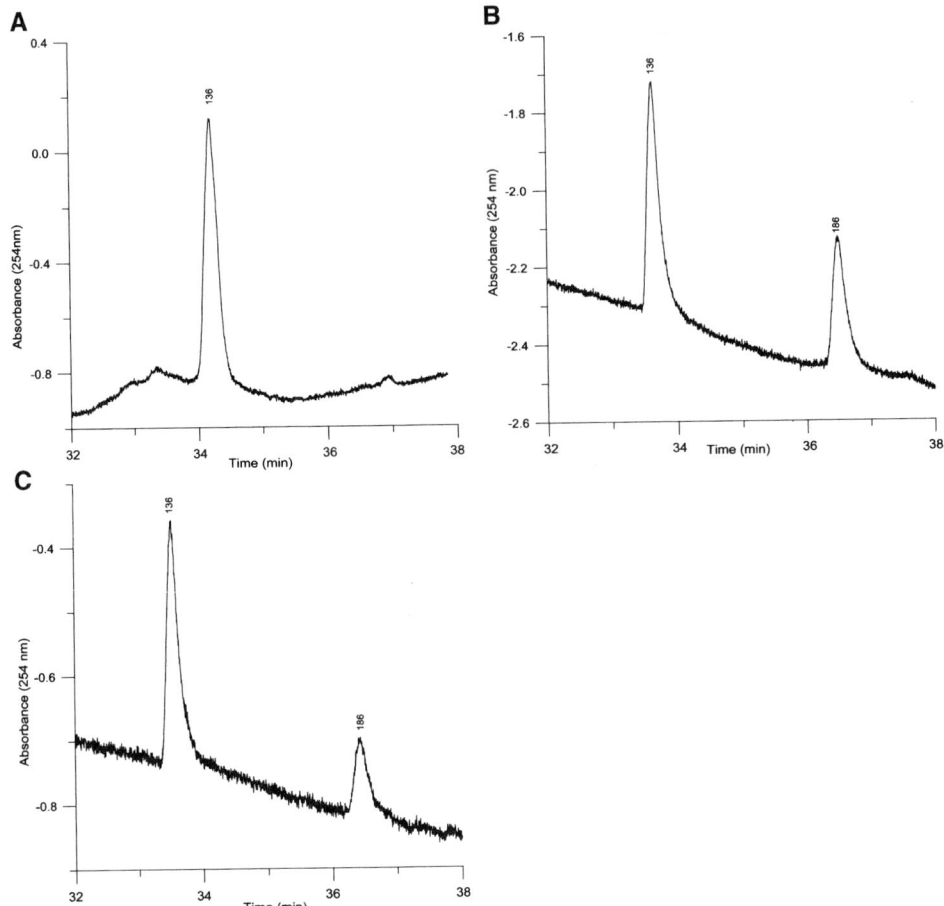

Fig. 3. CZE of PCR-amplified markers for Rh D assessment. Capillaries: 75 μm I.D, 37 cm length (30 cm to the detector), coated with poly(N-acryolyl amino propanol) *(41)*. Background electrolyte: 89 m*M* Tris-borate, 2 m*M* EDTA, pH 8.3, containing 8% short-chain, low-viscosity, liquid-sieving polyacrylamide. Sample injection: 15 s, at 100 V/cm. Run: 100 V/cm, room temperature. **(A)** Rh D-negative individual, exhibiting only the 136-bp peak; **(B)** Rh D-positive sample; because of the 2:1 peak ratio (136-bp/186-bp), it has been classified as a D/D homozygous individual; **(C)** Rh D-positive sample; because of the 3:1 peak ratio (136-bp/186-bp), it has been classified as a D/d heterozygous individual. (Reproduced with permission from **ref. *17*.**)

ity based on the quantitative expression ratio of the 136-bp/186-bp pair. Thus, a 2:1 peak ratio clearly indicates a D/D homozygous individual (**Fig. 3B**); a 3:1 peak ratio gives evidence of a D/d heterozygous individual (**Fig. 3C**).

4. Notes

1. Desalting of a PCR-amplified DNA fragment is a must for proper CZE analysis. Since the sample, because of the presence of a viscous polymer in the background electrolyte, cannot be injected by pressure, but only electrokinetically, salt in the sample will drastically reduce the amount of DNA injected in the capillary lumen.

2. Coated capillaries are necessary in order to drastically reduce the EOF flow, and thus allow better reproducibility of transit times from run to run. In addition, in the absence of coating, skewing of the DNA peaks can result. Coating the inner surface of a silica tube, via covalent attachment of polymeric material, is a complex job that requires skilled organic chemists; thus, the authors recommend buying commercially available capillaries, rather than making your own. In any case, an ample survey of coating procedures can be found in **ref. 36**.

3. Purging the capillary for long periods (e.g., overnight) with a viscous polymer solution (e.g., HEC, short-chain polyacrylamides) often results in a dynamic coating procedure, which can reduce the EOF flow to acceptable levels in uncoated capillaries. One could attempt using such procedure, and then normalize the transit times with the help of a marker of EOF flow. The authors use acrylamide (10 mM solution) as a marker of EOF *(37)*; others recommend mesityl oxide *(38)*.

4. Sieving liquid polymers in the CZE buffer are necessary in order to produce separation of DNA fragments of different size. However, the charge-to-mass ratio of DNAs does not plateau till reaching a length of 400 bp, above which all DNAs, in TAE buffer, exhibit a constant free mobility value of 3.75×10^{-4} cm^2/V/s^{-1} (and of 4.10×10^{-4} cm^2/V/s^{-1} in TBE; *33*). Thus, in principle, one could attempt separation of short DNA fragments in free solution. In practice, though, the relatively high diffusion coefficient of such short segments would deteriorate the separation.

5. It is possible to substitute short-chain liquid polyacrylamide with hydroxyethyl cellulose (HEC), which is commercially available. It is a highly stable polymer, and exerts proper sieving of even short DNA fragments. The authors recommend the 27 kDa size (number average molecular mass) from Polysciences, Warrington, PA in concentrations ranging from 3–6% (w/v). For more details, *see* Gelfi et al. *(34)*. For additional insight on the use of different liquid polymers in DNA separations, *see* **ref. 35**.

6. Do not attempt to perform separations in a true gel phase, such as a crosslinked polyacrylamide gel, polymerized in the capillary lumen. Although, in principle, a chemical gel exerts superior sieving and offers better resolution, in practice these gels quickly deteriorate because of the trapping of air bubbles and shrinking of gel extremities. Sieving liquid polymers, especially those of relatively lower viscosities, represent a flowable filling, and can be easily replaced at the end of each run, thus restoring identical initial conditions at each subsequent run.

7. Ethidium bromide (EtBr, a classical intercalator of ds-DNA) can be added to the background electrolyte in order to further improve the separation by modulating the DNA mobility (it preferentially binds to GC-rich regions, and stiffens the double helix). The concentration level is quite critical, though, since an excess of EtBr degrades the separation; the authors typically recommend a level of 2–3 µM *(40)*.

8. Those more skilled in CZE analysis, and desiring to avoid sample desalting, can attempt on-line isotachophoretic sample preconcentration by using chloride as leading ion and butyrate as terminator *(39)*. By this method, up to 700 nL of sample can be injected, as opposed to 5–8 nL by conventional injection procedures.
9. Laser-induced fluorescence detection increments the sensitivity by as much as $10^4 \times$ compared with UV detection. Under these conditions, no sample preconcentration or desalting is required.
10. When using plain UV detection based on intrinsic absorbance of DNA, a single round of amplification is often inadequate. For the procedures outlined in **Subheadings 3.1.** and **3.2.**, the authors recommend five rounds of amplifications, pooling the five different PCRs, and then proceeding to desalting and concentration, as detailed in **Subheading 3.2.1.**

Acknowledgments

Supported in part by grants from the European Community, Human Genome Project (Nos. BMH4-CT97–2627 and BMH4-CT96–1158), from Telethon-Italy (No. E.555), and from AIRC (associazione Italiana Ricerca sul cancro).

References

1. Landers, J. P. (ed.) (1997) Capillary electrophoresis in the clinical sciences. *Electrophoresis* **18,** 1709–1905.
2. Krstulovic, A. M. (ed.) (1997) Capillary electrophoresis in the life sciences. *J. Chromatogr. B.* **697,** 1–293.
3. Righetti, P. G. and Gelfi, C. (1997) Capillary electrophoresis of DNA for molecular diagnostics. *Electrophoresis* **18,** 1709–1714.
4. Wang, Y., Hung, S. C., Linn, J. F., Steiner, G., Glazer, A. N., Sidransky, D., and Mathies, R. A. (1997) Microsatellite-based cancer detection using capillary array electrophoresis and energy transfer fluorescent primers. *Electrophoresis* **18,** 1742–1749.
5. Personnet, D., Sugaya, K., Hammond, D., Robbins, M., and McKinney, M. (1997) Use of capillary electrophoresis with LIF-detection to assess mRNA molecules amplified by PCR: applications in the cloning of the cells. *Electrophoresis* **18,** 1750–1759.
6. Righetti, P. G. and Gelfi, C. (1997) Non-isocratic capillary electrophoresis for the detection of DNA point mutations. *J. Chromatogr. B.* **697,** 195–206.
7. El Rassi, Z. (ed.) (1997) Capillary electrophoresis and electrochromatography. *Electrophoresis* **18,** 2123–2502.
8. Righetti, P. G. and Gelfi, C. (1996) Capillary electrophoresis of DNA, in *Capillary Electrophoresis in Analytical Biotechnology* (Righetti, P. G., ed.), CRC, Boca Raton, FL, pp. 431–476.
9. Righetti, P. G. and Gelfi, C. (1997) Recent advances in capillary electrophoresis of DNA fragments and PCR products in poly(N-substituted acrylamides). *Anal. Biochem.* **244,** 95–207.
10. Righetti, P. G. and Gelfi, C. (1997) in *Analysis of nucleic acids by capillary electrophoresis* (Heller, C., ed.), Vieweg, Braunschweig, Germany, pp. 255–273.

11. Righetti, P. G. and Gelfi, C. (1997) Recent advances in capillary electrophoresis of DNA fragments and PCR products. *Biochem. Soc. Trans.* **25,** 267–273.
12. Lehmann, R., Liebich, H. M., and Voelter, W. (1996) Application of capillary electrophoresis in clinical chemistry. Developments from preliminary trials to routine analysis. *J. Cap. Electrophor.* **3,** 89–110.
13. Baba, Y. (1996) Analysis of disease-causing genes and DNA-based drugs by capillary electrophoresis. Towards DNA diagnosis and gene therapy for human diseases. *J. Chromatogr. B.* **687,** 271–302.
14. Gelfi, C., Righetti, P. G., Brancolini, V., Cremonesi, L., and Ferrari, M. (1994) Capillary electrophoresis in polymer networks for analysis of PCR products: detection of F508 mutation in cystic fibrosis. *Clin. Chem.* **40,** 1603–1605.
15. Gelfi, C., Orsi, A., Leoncini, F., Righetti, P. G., Spiga, I., Carrera, P., and Ferrari, M. (1995) Amplification of 18 dystrophin gene exons in DMD/BMD patients: simultaneous resolution by capillary electrophoresis in sieving liquid polymers. *BioTechniques* **19,** 254–263.
16. Gelfi, C., Cossu, G., Carta, P., Serra, M., and Righetti P. G. (1995) Gene dosage in capillary electrophoresis: pre-natal diagnosis of Down's syndrome. *J. Chromatogr. A.* **718,** 405–412.
17. Cossu, G., Angius, A., Gelfi, C., and Righetti, P. G. (1996) Rh D/d genotyping by quantitative polymerase chain reaction and capillary zone electrophoresis. *Electrophoresis* **17,** 1911–1915.
18. Kuypers, A. W. H. M., Meijerink, J. P. P., Smetsers, T. F. C. M., Linssen, P. C. M., and Mensink, E. J. B. M. (1994) Quantitative analysis of DNA aberrations amplified by competitive PCR using capillary electrophoresis. *J. Chromatogr. B.* **660,** 271–277.
19. Felmlee, T. A., Mitchell P. S., Ulfelder, K. J., Persing, D. H., and Landers, J. P. (1995) Capillary electrophoresis for the post-amplification detection of a hepatitis C virus-specific DNA products in human serum. *J. Cap. Electrophor.* **3,** 125–130.
20. Rossomando, E. F., White, L., and Ulfelder, K. J. (1994) Capillary electrophoresis: separation and quantitation of reverse transcriptase PCR products from polio virus. *J. Chromatogr. B.* **656,** 159–168.
21. Lu, W., Han, D. S., Yuan, and, J., Andrieu (1994) Multi-target PCR-analysis by capillary electrophoresis and laser-induced fluorescence. *Nature* **368,** 269–271.
22. Kolesar, J. M., Allen, P. G., and Doran, C. M. (1997) Direct quantification of HIV-1 RNA by capillary electrophoresis with laser-induced fluorescence. *J. Chromatogr. B.* **697,** 189–194.
23. Williams, S. J., Schwer, C., Krishnarao, A. S. M., Heid, C., Karger, B. L., and Williams, P. M. (1996) Quantitative competitive PCR: analysis of amplified products of the HIV-1 gag gene by capillary electrophoresis with LIF detection. *Anal. Biochem.* **236,** 146–152.
24. Gelfi, C., Leoncini, F., Righetti, P. G., Cremonesi, L., di Blasio, A. M., Carniti, C., and Vignali, M. (1995) Separation and quantitation of reverse transcriptase polymerase chain reaction fragments of basic fibroblast growth factor by capillary electrophoresis in polymer networks. *Electrophoresis* **16,** 780–783.

25. Lubin, M. B., Elashoff, J. D., Wang, S. W., Rotter, J. T., and Toyoda, H. (1991) Quantitative PCR for X-chromosome dosage in patients with sex chromosomal aneuploidies. *Mol. Cell Probes* **5,** 307–312.
26. Mansfield, E. S. (1993) Detection of trisomies 21 and 18 and of the triple-X syndrome by quantitative PCR. *Hum. Mol. Genet.* **2,** 43–50.
27. Bennett, P. R., Le Van Kim, C., Colin, Y., Warwick, R. M., Path, F. R. C., Chërif-Zahar, B., Fisk, N. M., and Cartron, J. P. (1993) Prenatal determination of fetal Rh D type by DNA amplification. *N. Engl. J. Med.* **329,** 607–610.
28. Chërif-Zahar, B., Raynal, V., Le Van Kim, C., D'Ambrosio, A. M., Bailly, P., Cartron, J. P., and Colin, Y. (1993) Structure and expression of the Rh locus in the Rh-deficiency syndrome. *Blood* **82,** 656–662.
29. Sherma, V. and Litt, M. (1992) Highly polymorphic, chromosome 21-specific D21S11 marker for Down's syndrome. *Hum. Mol. Genet.* **1,** 67–73.
30. Gelfi, C., Orsi, A., Leoncini, F., and Righetti, P. G. (1995) Fluidified polyacrylamides as molecular sieves in capillary zone electrophoresis of DNA fragments. *J. Chromatogr. A.* **689,** 97–105.
31. Krowczynska, A. M., Donoghue, K., and Hughes, L. (1995) Recovery of DNA, RNA and proteins from gels with microconcentrators. *BioTechniques* **18,** 698–703.
32. Williams, P. E., Marino, M. A., Del Rio, S. A., Turni, L. A., and Devaney, J. M. (1994) Analysis of DNA restriction fragments and PCR products by capillary electrophoresis. *J. Chromatogr. A.* **680,** 525–540.
33. Stellwagen, N., Gelfi, C., and Righetti, P. G. (1997) Free solution mobility of DNA. *Biopolymers* **42,** 687–703.
34. Gelfi, C., Perego, M., Libbra, F., and Righetti, P. G. (1996) Comparison of the behaviour of N-substituted acrylamides and celluloses on double-stranded DNA separations by capillary electrophoresis at 25° and 60°C. *Electrophoresis* **17,** 1342–1347.
35. Sunada, W. M. and Blanch, H. W. (1997) Polymeric separation media for capillary electrophoresis of nucleic acids. *Electrophoresis* **18,** 2243–2254.
36. Chiari, M., Nesi, M., and Righetti, P. G. (1996) Surface modification of silica walls: a review of different methodologies, in *Capillary Electrophoresis in Analytical Biotechnology* (Righetti, P. G., ed.), CRC, Boca Raton, FL, pp. 1–36.
37. Ermakov, S. V., Capelli, L., and Righetti, P. G. (1996) Method for measuring very weak, residual electroosmotic flow in coated capillaries. *J. Chromatogr. A.* **744,** 55–62.
38. Grossman, P. D. and Soane, D. S. (1990) Orientation effects on the electrophoretic mobility of rod shaped molecules in free solution. *Anal. Chem.* **62,** 1592–1599.
39. Van der Schanz, M. J., Beckers, J. L., Molling, M. C., and Everaerts, F. M. (1995) Intrinsic isotachophoretic preconcentration in capillary gel electrophoresis of DNA fragments. *J. Chromatogr. A.* **717,** 139–147.
40. McCord, B. R., McClure, D. L., and Jung, J. M. (1993) CZE of polymerase chain reaction amplified DNA using detection with an intercalating dye. *J. Chromatogr.* **652,** 75–82.
41. Simo-Alfonso, E., Gelfi, C., Sebastiano, R., Citterio, A., and Righetti, P. G. (1996) Novel acrylamido monomers with higher hydrophilicity and improved hydrolytic stability. II: properties of N-acryloyl amino propanol. *Electrophoresis* **17,** 732–737.

13

Rapid Analysis of Amplified Double-Stranded DNA by Capillary Electrophoresis with Laser-Induced Fluorescence Detection

Ming-Sun Liu and Fu-Tai Albert Chen

1. Introduction

Electrophoresis is an important and fundamental tool for DNA analysis. Traditional DNA electrophoresis is labor-intensive, skill-dependent, and relatively slow. To achieve greater performance, capillary electrophoresis (CE) has been developed for DNA work *(1–2)*. The DNA fragments can be separated by molecular sieving in CE with gel-filled capillaries, based on the same principle as that of the conventional gel electrophoresis. The whole procedure can be completed in a relatively short period with great resolution. Recently, CE equipped with laser-induced fluorescence (CE-LIF) has been used as a tool for DNA analysis *(3)*. Using fluorescence DNA intercalators in a gel-filled capillary, the detection sensitivity was found to increase 2–3 orders of magnitude over that of UV detection, and reaches picogram levels *(3)*.

Polymerase chain reaction (PCR) has been widely used for clinical diagnostics and research *(4)*. However, separation and detection of PCR-amplified double-stranded DNA (ds = DNA) are commonly based on agarose gel electrophoresis in combination with ethidium bromide (EB) staining. PCR-amplified DNA fragments can also be separated by molecular sieving in CE with polymer-filled capillaries. Using a linear polyacrylamide gel-filled capillary for DNA analysis in combination with LIF, a single run can be completed within 15–30 min *(5)*. However, for clinical detection involving large numbers of samples, 15-min electrophoresis time for each assay is too long.

This chapter demonstrates the rapid analysis of amplified ds = DNA fragments by CE-LIF system using high-pH buffer and EB as a DNA intercalator,

From: *Methods in Molecular Medicine, Vol 27: Clinical Applications of Capillary Electrophoresis*
Edited by: S. M. Palfrey © Humana Press Inc., Totowa, NJ

without the addition of sieving materials. The fast separation is based on the slight differences in charge-to-mass ratio of different sizes of ds = DNA fragments. The migration of ds = DNA by CE is driven by electroosmotic flow in the fused-silica capillary. The smaller ds = DNA fragments have smaller electrophoretic mobility, and can be driven fast by electroosmotic flow and appear first through the detection window. It is a 3-min rapid analysis vs a 15–20 min of the sieving-based separation. More than 10 samples can be automatically analyzed by CE within 1 h, and the data is entered directly into a computer. This method can be used in a clinical laboratory for large-scale PCR-amplified samples analysis.

2. Materials

2.1. Hardware

1. CE unit: P/ACE™ 2100 system (Beckman, Fullerton, CA).
2. Laser induced fluorescence detector (Beckman).
3. A 5-mW He/Ne laser source, excitation at 543 nm and emission at 600 nm (Melles Griots, Irvine, CA).
4. Bare fused silica capillary (22 µm id × 27 cm) (Polymicro Technology, Phoenix, AZ).
5. IBM-compatible personal computer with Microsoft Windows™ 3.X and P/ACE™ 3.0 software.

2.2. Solutions

2.2.1. PCR Amplification Solutions for Human ZFY Gene

1. 0.5-mL plastic tube (100 µL final vol/reaction).
2. 80 µL deionized water.
3. 10 µL 10X reaction buffer: 100 mM Tris-HCl, 15 mM MgCl$_2$, 500 mM KCl, pH 8.3 (Boehringer Mannheim, Indianapolis, IN).
4. 0.8 µL 25 mM dNTP (25 mM each dATP, dTTP, dCTP, and dGTP).
5. 0.5 µL Taq DNA polymerase (10 U/µL).
6. 1 µL primer 5' and primer 3' 20 pM (5'-CTCTCAGTTCACACAAAGG-3' and 5'-GCTTGTAGACACACTGTTAGG-3') for ZFY gene amplification.
7. 1 µL primer 5' and primer 3' 20 pM(5'-ACAGAATTCGCCCCGGCCTGGTACAC-3') and (5'-TAAGCTTGGCACGGCTGTCCAAGGA-3') for human ApoE gene amplification.
8. This makes up a volume of 99 µL, and is ready for the addition of 1 µL human genomic DNA (1 µg/µL).

2.2.2. Buffers for Capillary Electrophoresis

1. Borate buffer: 100 mM borate buffer, pH 10.0, containing 10 µM ethidium bromide (EB).
2. Wash: 1 M NaOH.
3. 5 mL deionized water.

3. Methods
3.1. PCR Amplification

1. Extracted genomic DNA template for PCR amplification was from peripheral blood of a healthy human male. Prepare oligonucleotide primers for PCR amplification on the OLIGO 1000 DNA synthesizer (Beckman).
2. Obtain a 307-bp amplified DNA fragment from amplified human ZFY gene, according to the protocols described by Pao et al. (6). Briefly, the 100 µL amplification reaction mixture contains 1 µg human male genomic DNA, 20 pM each of the two primers, 20 nM each of four deoxyribonucleoside triphosphates, and 5 U of Taq DNA polymerase (Boehringer Mannheim). Carry out the reaction in 0.5-mL microcentrifuge tubes, and overlay with mineral oil.
3. Use an initial denaturation at 94°C for 5 min, followed by 30 cycles of template denaturation at 94°C for 1 min, primer-template annealing at 55°C for 1 min, and primer extension at 72°C for 1 min. At the end of cycles, add a 10-min incubation at 72°C .
4. Perform DNA amplification of the ApoE gene, according to the protocols described previously (7). Briefly, the 30 µL amplification reaction mixture contains 1 µg human male genomic DNA, 10% dimethyl sulfoxide, 30 pM each of the two primers, 2.5 mM each of four deoxyribonucleoside triphosphates, and 5 U of Taq DNA polymerase (Boehringer Mannheim). Carry out the reaction in 0.5-mL microcentrifuge tubes, and overlay with mineral oil.
5. Heat the sample at 95°C for 5 min, followed by 30 cycles of template denaturation at 95°C for 1 min, primer-template annealing at 60°C for 1 min, and primer extension at 70°C for 2 min. At the end of the PCR cycles, add a 10-min incubation at 72°C. A 244-bp DNA fragment is obtained.

3.2. Capillary Electrophoresis

In the authors' studies, the P/ACE system 2100 CE, equipped with LIF detection (Beckman), was utilized as standard equipment.

1. The anode is set on the injection side, and the cathode is set on the detection side. Two 4.5-mL bottles, filled with borate buffer containing EB, are placed in inlet and outlet tray positions 11 and 1 as electrode buffer.
2. A 4.5-mL bottle filled with deionic water is placed on inlet tray position 34, and a bottle filled with 1 M NaOH is placed on inlet tray position 33.
3. Undiluted PCR amplified ds = DNA solution (1–3 mg/L) is placed in sample tray position 15 (*see* **Note 1**).
4. A 5 mW He/Ne laser source (Melles Griots) associated with band-pass filters (excitation at 543 nm and emission at 600 nm) was set for detection of EB-DNA complex.
5. Assemble a 22 µm id × 27 cm total length (20 cm to detector) bare fused-silica capillary column in a P/ACE LIF cartridge. The separation temperature is set at 22°C. Rinse the capillary (20 psi) with 1 M NaOH for 5 min and deionic water for 5 min (*see* **Note 2**).

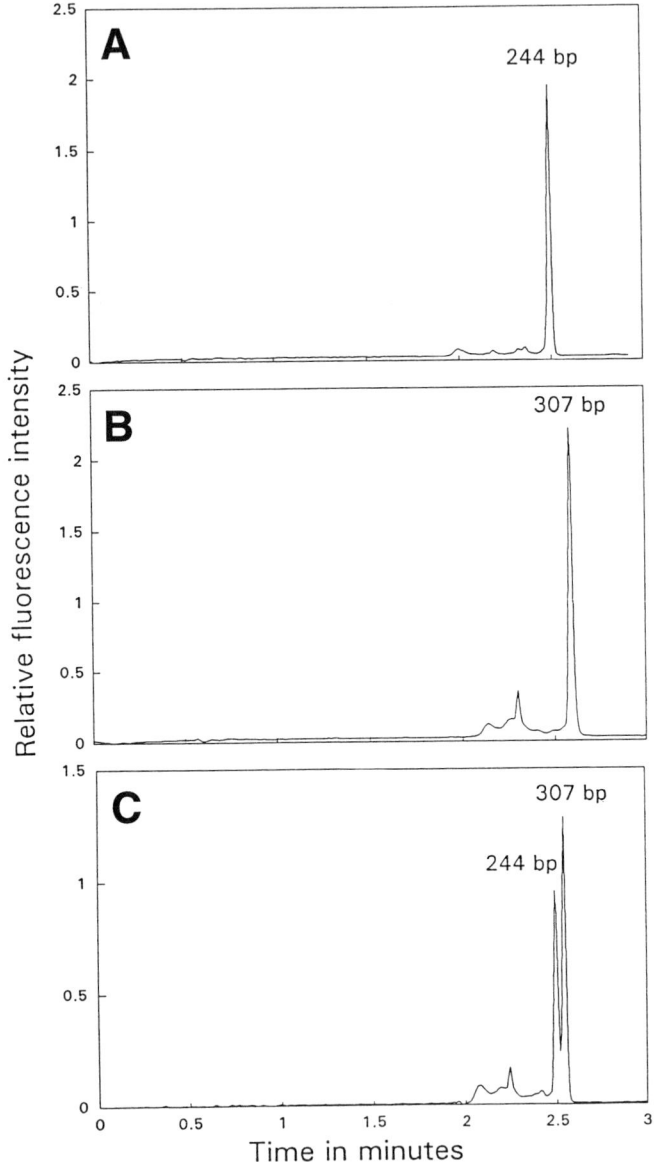

Fig. 1. Electropherograms of PCR amplified DNA fragments by CE-LIF. (**A**) Amplified DNA fragments of 244 bp in size. (**B**) Amplified DNA fragments of 307 bp in size. (**C**) Mixture of DNA fragments of 244 and 307 bp in size.

6. The separation procedure is listed in **Fig. 1**. Rinse the capillary with borate buffer for 30 s, and PCR-amplified DNA sample is directly injected into the capillary by

pressure (0.5 psi) for 20 s. Separation is achieved at a field strength of 1000 V/cm for 3 min, and the electropherograms are recorded using P/ACE 3.0 (**Fig. 1**; *see* **Note 3**). A high-performance (20 psi) rinse of 1 M NaOH for 5 s, followed by water for 30 s, is used between runs (*see* **Note 4**).

7. The advantages of using CE-LIF for PCR-based clinical diagnosis are as follows: First, CE-LIF analysis only consumes a few nanoliters of DNA solution. Second, CE-LIF provides very rapid and sensitive detection. Third, the data could be directly entered into the computer for further analysis. Furthermore the present method can also been used in a multicapillary-based CE system to achieve maximal throughput performance.

8. In conclusion, the CE-LIF system provides a great potential for the future DNA analysis, especially for DNA amplification-based genetic diagnosis.

4. Notes

1. A 10-μL vol of PCR-amplified DNA solution is needed in to sample vial for sample injection.

2. In this demonstration, the CE-LIF system provides a simple, rapid, sensitive, and automated method to analyze PCR-amplified DNA fragments. A bare fused capillary column may be used for thousands of runs in the CE-LIF system without damage. However, the detection window of the capillary is its most fragile part of, and must be handled with care to avoid breaking.

3. According to the authors' calculation, a volume of 1.6 nL of DNA solution was injected into the capillary in each analysis. It represented that approx 10 pg of amplified DNA was introduced into the capillary. In comparison, the detection sensitivity is about 4–5 orders of magnitude over that by agarose gel analysis using UV detection.

4. The operation and maintenance of the P/ACE™ 2100 system should follow the Beckman P/ACE™ series instrument manual.

Acknowledgment

We thank Dr. James C. Osborne for his support.

References

1. Strege, M. and Lagu, A. (1991) Separation of DNA restriction fragments by capillary electrophoresis using coated fused silica capillaries. *Anal. Chem.* **63,** 1233–1236.
2. Ulfelder, K. J., Schwartz, H. E., Hall, J. M., and Sunnzeri, F. J. (1992) Restriction fragment length polymorphism analysis of ERBB2 oncogene by capillary electrophoresis. *Anal. Biochem.* **200,** 260–267.
3. Schwartz, H. E. and Ulfelder, K. J. (1992) Capillary electrophoresis with laser induced fluorescence detection of PCR fragments thiazole orange. *Anal. Chem.* **64,** 1737–1740.
4. Rolfs, A., Schuller, L., Finckh, U., Weber-Rolfs, I., eds. (1992) *PCR: Clinical Diagnostics and Research.* Springer-Verlag, Berlin.

5. Liu, M. S., Zang, J., Evangelista, R. A., Rampal, S., and Chen, F. T. A. (1995) Double-stranded DNA analysis by capillary electrophoresis with laser-induced fluorescence using ethidium bromide as an intercalator. *BioTechniques* **18,** 316–323.
6. Pao, C. C., Kao, S. M., Hor, J. J., and Chang, S. Y. (1993) Lack of mutational alteration in the conserved regions of ZFY and SRY gene of 46, XY females with gonadal dysgenesis. *Hum. Reprod.* **8,** 224–228.
7. Hixson, J. E. and Vernier, D. T. (1990) Restriction isotyping of human apolipoprotein E by gene amplification and cleavage with HhaI. *J. Lipid Res.* **31,** 545–548.

14

Identification of Mutated p53 in Cancers by Nongel-Sieving Capillary Electrophoretic SSCP Analysis

Michiei Oto

1. Introduction

The tumor suppressor gene, p53, lies on chromosome 17p, and has been examined in a wide variety of primary tumors, xenografts, and cell lines derived from tumors. Point mutations in the evolutionally conserved codons of p53 have appeared to be the most common genetic alterations in human cancers *(1)*. The p53 mutational spectrum differs among those of cancers of the colon, lungs, esophagus, breast, liver, brain, reticuloendothelial tissues, and hemo-phoietic tissues. In particular, 75–80% of colon carcinomas exhibit a loss of both p53 alleles, one through deletion and the other through point mutation *(1)*.

To study the point mutation in the p53 gene, it is necessary to develop rapid and easy analytical techniques for nucleotide sequences. There are several analytical techniques for detecting a single-point mutation, for example, single-strand conformation polymorphism (SSCP) analysis *(2,3)*, sequencing *(4)*, denaturing gradient gel electrophoresis (DGGE) analysis *(5)*, hybridization with a point mutation specific oligonucleotide probe *(6)*, polymerase chain reaction-restriction fragment length polymorphism (PCR-RFLP) analysis *(7)*, and RNase protection analysis *(8)*.

SSCP analysis is popular because of its simplicity, and its wide availability and applicability to various genes *(2,3)*. In this technique, a PCR-amplified double-stranded (ds) DNA is denatured into a single-stranded (ss) DNA, and then the conformationally different denatured DNA corresponding to the nucleotide sequences is separated by polyacrylamide gel electrophoresis. Electrophoretic mobility shifts indicate differences in the nucleotide sequences of interest. Since, at low concentrations (pg-fg/mL), the renaturation of ssDNA

From: *Methods in Molecular Medicine, Vol 27: Clinical Applications of Capillary Electrophoresis*
Edited by: S. M. Palfrey © Humana Press Inc., Totowa, NJ

rarely occurs and the proportion of ssDNA is dominant, ssDNA has been labeled with radioactive materials, including ^{32}P, or stained with silver. To determine a mutant sequence, the mutant DNA must be purified from the gel, and subjected to cycle sequencing, or cloned by the thymine and adenine (TA) cloning technique *(9)* for sequencing.

In contrast to gel electrophoresis, capillary electrophoresis (CE) has been reported as a promising technique for the rapid and reproducible separation of dsDNA, as well as ssDNA, and its use is feasible for automated DNA analysis *(10)*. In addition, the separated DNA can be fractionated by CE, which is very useful, for example, for direct sequencing. Nongel sieving CE is a specialized form of capillary zone electrophoresis performed with a buffer containing polymer additives, such as methylcellulose, and is easier to perform than gel-sieving CE *(11)*. These additives change the rate of migration, depending on the molecular mass or the molecular conformation. A nongel-sieving capillary can only be prepared by introducing a buffer containing polymer additives into a capillary, and this technique gives reproducible results, and the capillary is reusable *(11,12)*. In electrophoresis, DNA is monitored with a UV detector or a LIF detector, and the acquisition of data, including their analyses, is computer-controlled, which facilitates automated DNA detection in multiple samples, without a staining procedure.

This technique can allow the separation of single-stranded, as well as double-stranded, DNA in various size ranges within 30–40 min, under the optimal conditions; hence, the nongel-sieving CE technique is useful for the analysis of DNA, including PCR products, and has been applied to various DNA diagnostic techniques *(12–15)*, including the SSCP technique *(13,16)*.

In this chapter, the detection of a point mutation in exons 5–6 of the p53 gene, derived from cancer specimens by means of the capillary electrophoretic SSCP technique, is described. This method involves a system for detecting a point mutation, including SSCP analysis with nongel-sieving CE, followed by the collection and sequencing of the mutant DNA. In this technique, there are some critical factors, which include the temperature for the electrophoresis and the efficiency of the collection of DNA by CE. These factors are also discussed, using data or figures, in this chapter.

2. Materials

2.1. Reagents and Instrumentation for the PCR Technique

1. Cyclone Plus (Millipore, Bedford, MA) for the preparation of oligonucleotide primers.
2. Oligo Pak (Millipore) for the purification of oligonucleotide primers.
3. Sterilized distilled water for dissolving the primers.
4. Zymoreactor II incubator (ATTO, Tokyo, Japan) to carry out DNA amplification.

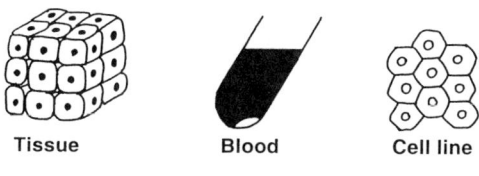

DNA extraction
⬇
Gene amplification by PCR
⬇
Denaturation of PCR products
⬇
CE analysis
⬇
DNA collection by CE
⬇
Cycle sequencing analysis
⬇
Detection of mutant sequence

Fig. 1. Summary of CE-SSCP analysis.

5. Deoxyribonucleotide triphosphates (Toyobo, Osaka, Japan).
6. Taq DNA polymerase (Toyobo).
7. Reaction buffer: 10 mM Tris-HCl, pH 8.8, 50 mM KCl, 1.5 mM MgCl^2, and 0.1% gelatin in a reaction tube.
8. Reaction tubes (Robbins Scientific, Sunnyvale, CA) suitable for the incubator.

2.2. Reagents and Instrumentation for Capillary Electrophoresis

1. CE was performed on a capillary column (Bio-Rad, Hercules, CA).
2. 3X TBE buffer: 0.267 M Tris/borate, pH 8.3, 1 mM EDTA containing polymer additives (PCR Product Analysis Buffer™; Bio-Rad.).
3. Laser fluorescence detector: BioFocus™ 3000 LIF^2 (Bio-Rad).
4. Washing solution: (Capillary Washing Solution, Bio-Rad).
5. Chemiluminescence DNA cyclesequencing kit (Biotinylate Terminator Sequencing High Plus kit™, Toyobo).
6. Fluorescent dye cycle sequencing kit (PRISM™, Dye Terminatior cycle sequencing kit; Perkin-Elmer, Foster City, CA).
7. ABI 310 genetic analyzer (Perkin-Elmer) for CE sequencing.

3. Methods

A summary of the capillary electrophoretic SSCP (CE-SSCP) analysis is shown in **Fig. 1.** The extracted DNA was amplified by the PCR technique, and then subjected to CE after denaturation. The single-stranded DNA peaks of interest were collected, and then subjected to cycle sequencing to determine the point mutations.

3.1. DNA Templates

Surgical specimens were obtained from colorectal cancer patients, according to the procedures of the Helsinki Declaration of 1975, as revised in 1983. High-mol-wt DNA from normal mucosa and tumor tissue of the same patient is extracted by a method described previously *(17)*.

3.2. DNA Amplification of Exons 5–6 of the p53 Gene

1. The forward primer (P1) sequence is 5'-TTCCTCTTCCTGCAGTACTC-3'.
2. The reverse primer (P2) sequence is 5'-GCAAATTTCCTTCCACTCGG-3'.
3. Mix 500 ng genomic DNA, 500 nM oligonucleotide primers, 20 mM dNTP, 0.5 U Taq DNA polymerase, in 10 mM Tris-HCl, pH 8.8, 50 mM KCl, 1.5 mM MgCl2, and 0.1% gelatin in a reaction tube.
4. The total volume of this reaction mixture is 50 μL.
5. Perform 40 cycles of reaction at 94°C for 1 min, 63°C (*see* **Note 1**) for 2 min, and 72°C for 1 min in a program incubator.

3.3. Denaturation of PCR Products

Denature the PCR products at 94°C for 5 min, and then chill on ice for 1 min, before subjecting them to nongel CE directly without further purification.

3.4. Capillary Electrophoretic Analysis

1. Immediately after denaturation, subject the PCR products to CE analysis.
2. Use a CE instrument with a UV detector (BioFocus 3000) or a laser-induced fluorescence detector (BioFocus 3000 LIF[2]). A summary of the CE conditions is given in **Fig. 2** (*see* **Note 6**).
3. Use a capillary cartridge containing a 50 μm (id) × 36 cm linear polyacrylamide-coated capillary.
4. Prior to sample migration, filter the washing solution containing detergent, distilled water, and the sieving buffer containing polymer additives, and degas the samples by centrifugation.
5. Flush the capillary for 30 s with the washing solution, and then for 30 s with distilled water, before each run.
6. After washing the capillary, introduce the buffer containing polymer additives into the capillary by means of a pressurized injection at 80 psi for 120 s (*see* **Note 2**).
7. Introduce the DNA sample by pressurized injection at 80 psi (10 nL of the DNA sample is loaded into the capillary for analysis).
8. Electrophorese DNA samples for 30 min at 139V/cm and 25°C (*see* **Notes 3** and **4**) in PCR Product Analysis Buffer.
9. Set the UV detector to 260 nm, with a range of 0.02 absorbance units.

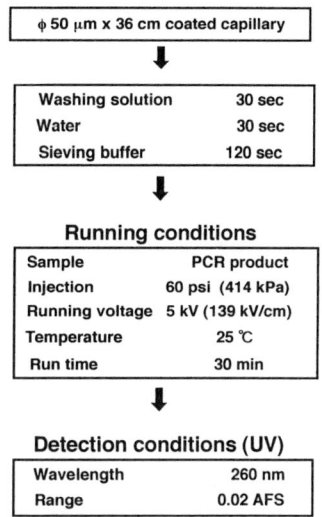

Fig. 2. CE–SSCP conditions for nongel-sieving CE.

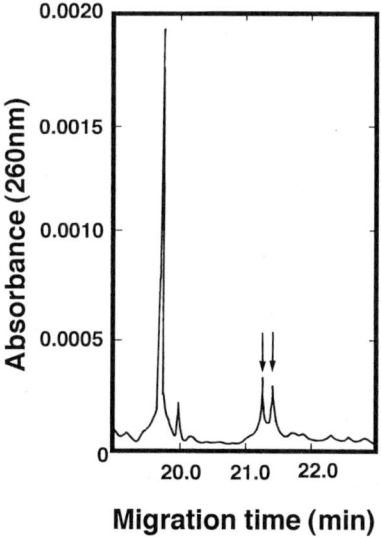

Migration time (min)

Fig. 3. Typical SSCP electropherogram. The main peak (migration time, 19.8 min) represents dsDNA of 325 bp, and the minor peaks (migration time, 21–22 min), indicated by arrows, are ssDNA with SSCP. The two peaks represent the plus and minus ssDNA strand.

10. After denaturation of the PCR products (325 bp), the plus and minus strands of ssDNA, which have specific conformations, separate from each other under these conditions (**Fig. 3**).

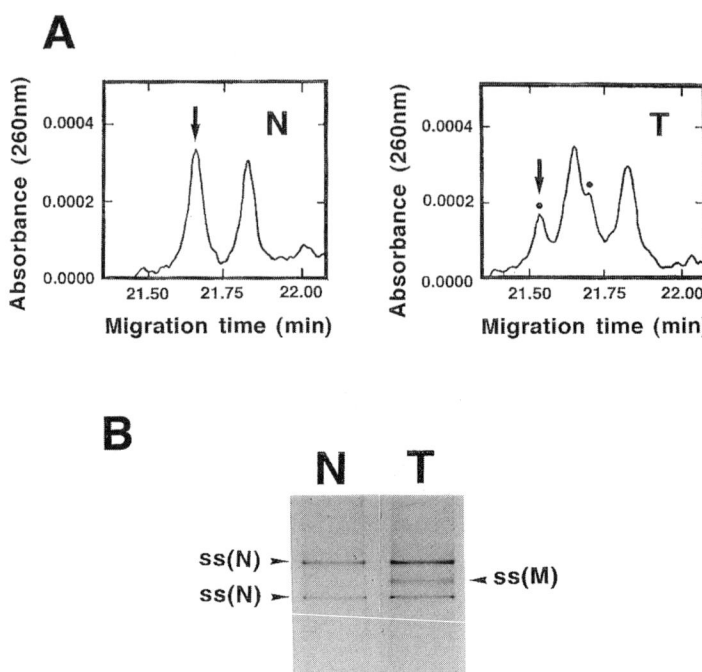

Fig. 4. SSCP analysis of the human p53 gene derived from normal DNA (N) and tumor DNA (T). (**A**) Nongel-sieving CE analysis and (**B**) minislab gel electrophoretic analysis are shown. The closed dots in (**A**) indicate additional peaks of tumor DNA, and the arrow indicates the peak collected for sequencing.

3.5. Typical Results of Detection of a Point Mutation Using CE-SSCP Analysis

On analysis of the normal DNA derived from a colorectal cancer patient by nongel-sieving CE, two SSCP peaks are observed in the electropherogram; for the tumor DNA of the same patient, additional SSCP peaks different from those for the normal DNA are detected within 30 min (**Fig. 4A**). These additional peaks are derived from the mutant DNA. The same results are obtained on SSCP analysis by minislab gel electrophoresis (**ref. *3*; Fig. 4B**; *see* **Note 5**).

3.6. Collection of Normal and Mutant ssDNA

1. To determine the DNA sequences of these materials, it is necessary to collect the first single-strand peaks of the normal and tumor DNA. After the first run, the time-range for collecting ssDNA is determined from the electropherogram, and then programmed with BioFocus Operating software before the second run *(13)*. In this experiment, the time-ranges for collection are 21.34–21.58 min (24 s) for a mutant peak and 21.52–21.74 min (24 s) for a normal peak (*see* **Note 6**).

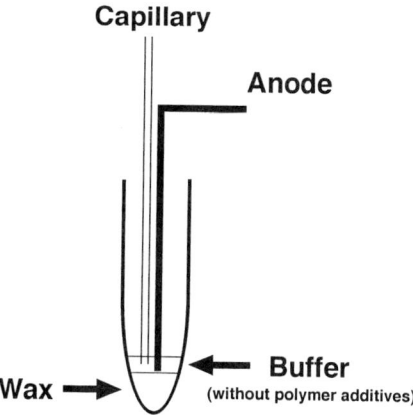

Fig. 5. Modification of vials for collection of ssDNA.

2. Collect the ssDNA in a vial containing 10 μL running buffer without polymer additives for 24 s.
3. Before the run, the vial used for collecting DNA should be modified, because the bottom of the vial is too narrow for a capillary or an electrode to reach. Add a piece of wax (AmpliWax™, Perkin-Elmer) to a vial, and then heat it to 70°C for 1 min to melt the wax. Allow it to cool to room temperature to solidify the wax. The shape of the inner bottom of the vial is changed, and the capillary and the anode can dip into the buffer (**Fig. 5**).

3.7. Cycle Sequencing Analysis of Collected ssDNA

1. The collected DNA (10 μL) is directly used as a template for cycle sequencing.
2. Use commercially available cycle sequencing kits (chemiluminescent detection, Toyobo; or fluorescence detection, Perkin-Elmer), according to their regular protocols.
3. **Figure 6** shows a comparison of conventional gel electrophoresis with CE, using a ABI310 genetic analyzer with a laser-induced fluorescence detector. In codon 175, CGC from the normal DNA, and CAC from the tumor DNA, were detected, respectively (**Fig. 6A,B**).

3.8. Advantages of This Method

These data demonstrate that nongel-sieving CE enables SSCP analysis within 30 min, and the collection of a mutant DNA for sequencing. To identify the point mutation, the system involving SSCP analysis, the collection of a mutant DNA by nongel-sieving CE, and cycle sequencing, described here, will be easier and more useful than conventional methods, because it is not necessary to prepare gels, to stain DNA, or to purify or clone the mutant DNA. Moreover, this method can be performed automatically in a short time. This system might be applicable to the DNA diagnosis of cancer.

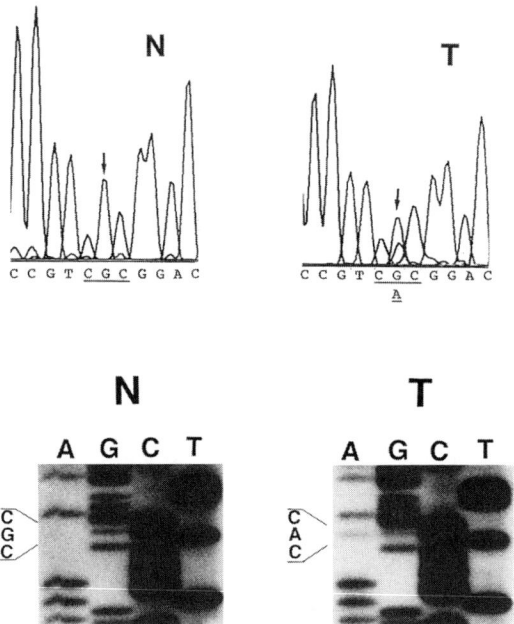

Fig. 6. Detection of mutations by cycle sequencing. (**A**) CE (ABI310) analysis and (**B**) conventional gel electrophoretic analysis are shown. In tumor DNA, the mutated sequences are indicated (CGC to CAC).

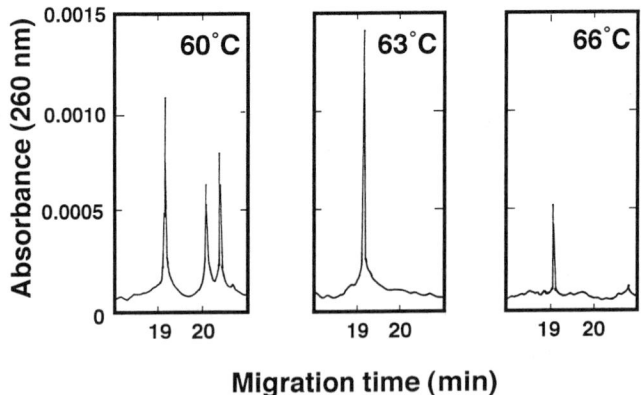

Fig. 7. Optimization of the PCR conditions. DNA derived from normal mucosae was amplified by the PCR technique, with an annealing temperature of 60, 63, or 66°C, and then subjected to CE analysis. The CE conditions were the same as in **Fig. 2**. In the electropherogram, the amplified 325-bp DNA fragment appeared at 19.1 min migration time.

4. Notes

1. The PCR cycle conditions (especially the annealing temperature) should be optimized for each target gene. On SSCP analysis, PCR products containing nonspecifically amplified DNA will show various and complex SSCP patterns, which are difficult to analyze. Therefore, PCR products should be evaluated by CE before being subjected to SSCP analysis. The capillary electrophoretic pattern of amplified p53 exon 5–6 DNA is shown in **Fig. 7**. These data demonstrate that annealing at 60, 63, and 66°C gave many nonspecifically amplified peaks, one specifically sharp peak, and one weak peak, respectively. It is clear that the optimal annealing temperature in this case is 63°C. Determination of the PCR products by CE should be performed under the same conditions (running temperature, voltage, and detector) as for the SSCP analysis.

2. Because a buffer containing polymer additives is highly viscous, the buffer containing polymer additives sometimes becomes attached to the surface of a capillary, depending on the concentration. To prevent carryover of the viscous buffer into the sample, it is desirable to dip the capillary into distilled water 3×.

3. On SSCP analysis, the ssDNA conformation will change, depending on the temperature. Therefore, the critical factors for SSCP analysis are the temperature and the voltage during electrophoresis. The stability of the two conformations of ssDNA (plus or minus strand) depends on the running temperature. In **Fig. 8**, *(shown on p. 136)* electropherograms of ssDNA under different running temperature conditions are presented. The conformations of the denatured PCR products which contained exons 5–6 of the p53 gene, were stable at 20–25°C and 139 V/cm. In **Fig. 9**, *(shown on p. 136)* electropherograms of ssDNA under different voltages are shown. A higher voltage during electrophoresis allows completion of the analysis in a shorter time, but brings about lower resolution of ssDNA, because of the higher temperature in the capillary. For exons 5–6 of the p53 gene, ssDNA should be electrophoresed at 25°C and 139 V/cm.

4. Since the stability of the conformation of ssDNA might depend on its sequence, the optimum conditions regarding temperature and voltage should be studied for each sequence of interest. The SSCP of PCR products (100–300 bp) might be detectable on electrophoresis at 15–30°C, according to these experiments. There are two types of methods for cooling a capillary, involving air-cooled and water-cooled systems, separately. But in an air-cooled system, the temperature in a capillary is not maintained at lower-than-room temperature. Therefore, CE instrumentation with a water-cooled system is recommended for SSCP analysis.

5. To confirm the differences between normal and cancer DNA, each DNA was amplified with primers labeled with FITC and Texas Red, respectively. Amplified DNA was subjected to CE under the same conditions as in **Fig. 2**, except the use of a laser-induced fluorescence (LIF) detector. The LIF conditions are shown in **Fig. 10** *(shown on p. 137)*. An electropherogram is shown in **Fig. 11** *(shown on p. 137)*. In this figure, a normal ssDNA and a mutant ssDNA were run on the same electropherogram, indicating the possibility of detecting small changes between them.

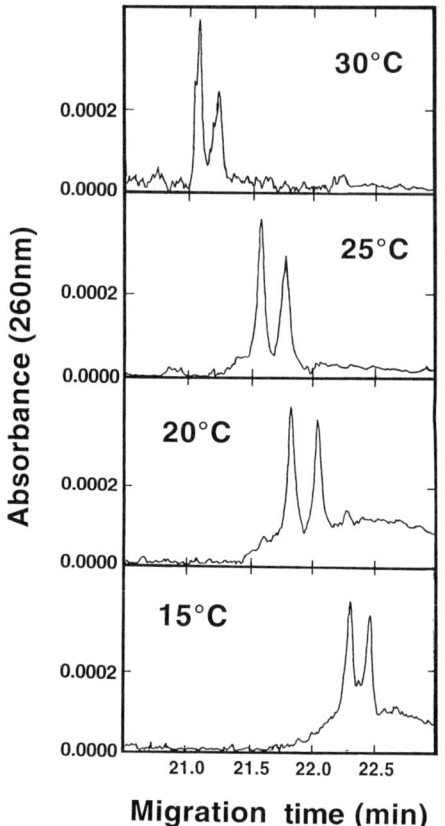

Fig. 8. Study of the running temperature. The CE conditions were the same as in **Fig. 2**, except for the running temperature.

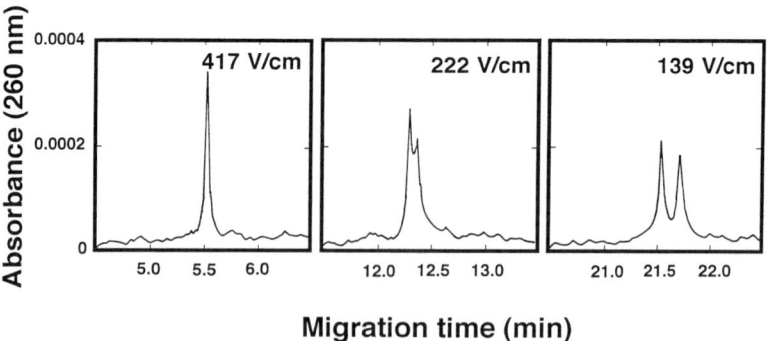

Fig. 9. Study of the running voltage. The CE conditions were the same as in **Fig. 2**, except for the running voltage.

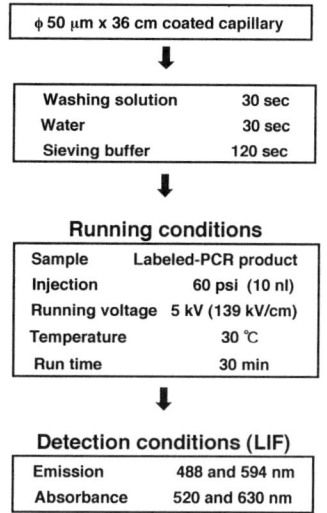

Fig. 10. SSCP analysis involving CE-LIF.

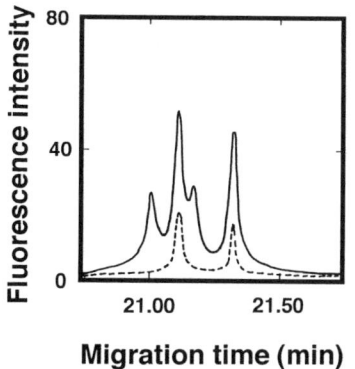

Migration time (min)

Fig. 11. Typical electropherograms of SSCP using CE-LIF. The differences in the profiles of normal and tumor DNA were determined with the double-injection technique. The electropherograms of DNA from normal and tumor tissues are shown by the dotted line and solid line, respectively.

6. The time-range should be broader, because the reproducibility (CV%) of the migration time is about 0.7–1.0%, and the amount of the collected ssDNA might be lower.

References

1. Levin, A. J., Momand, J., and Finlay, C. A. (1991) The p53 tumor suppressor gene. *Nature* **351,** 453–456.

2. Orita, M., Suzuki, Y., Sekiya, T., and Hayashi, K. (1989) Rapid and sensitive detection of point mutations and DNA polymorphisms using the polymerase chain reaction. *Genomics* **5,** 874–879.
3. Oto, M., Miyake, S., and Yuasa, Y. (1993) Optimization of nonradioisotopic single strand conformation polymorphism analysis with a conventional minislab gel electrophoresis apparatus. *Anal. Biochem.* **213,** 19–22.
4. Sanger, F., Nicklen, S., and Coulson, A. R. (1977) DNA sequencing with chain-terminating inhibitors. *Proc. Natl. Acad. Sci. USA* **74,** 5466–5467.
5. Fischer, S. G. and Lerman, L. S. (1983) DNA fragments differing by single base-pair substitutions are separated in denaturing gradient gels: correspondence with melting theory. *Proc. Natl. Acad. Sci. USA* **80,** 1579–1583.
6. Yuasa, Y., Oto, M., Sato, C., Miyaki, M., Iwama, T., Tonomura, A., and Namba, M. (1986) Colon carcinoma K-ras oncogene of a familial polyposis coli patient. *Jap. J. Cancer. Res.* **77,** 901–907.
7. Ando, M., Maruyama, M., Oto, M., Takemura, K., Endo, M., and Yuasa, Y. (1991) Higher frequency of point mutations in the c-k-ras^2 gene in human colorectal adenoma with severe atypia than in carcinoma. *Jpn. J. Cancer. Res.* **82,** 245–249.
8. Nishisho, I., Nakamura, Y., Miyoshi, Y., Miki, Y., Ando, H., Horii, A., et al. (1991) *Science* **253,** 665–669.
9. Marchuk, D., Drumm, M., Saulino, A., and Collins, F. S. (1991) Construction of T-vectors, a rapid and general system for direct cloning of unmodified PCR products. *Nucleic Acids Res.* **19,** 3154.
10. Martin, F., Vairelles, D., and Henrion, B. (1993) Automated ribosomal DNA fingerprinting by capillary electrophoresis of PCR products. *Anal. Biochem.* **14,** 182–189.
11. Zhu, M., Hansen, D. L., Burd, S., and Gannon, F. (1989) Factors affecting free-zone electrophoresis and isoelectric focusing in capillary electrophoresis. *J. Chromatogr.* **480,** 311–319.
12. Oto, M., Suehiro, T., Akiyama, Y., and Yuasa, M. (1995) Microsatellite instability in cancer identified by non-gel sieving capillary electrophoresis. *Clin. Chem.* **41,** 482–483.
13. Oto, M., Suehiro, T., and Yuasa, Y. (1995) Identification of mutated p53 in cancer by non-gel-sieving capillary electrophoretic SSCP analysis. *Clin. Chem.* **41,** 1787–1788.
14. Oto, M., Suehiro, T., and Yuasa, Y. (1995) DNA hybridization analysis of PCR products by non-gel sieving capillary electrophoresis. *PCR Methods. Applic.* **4,** 303–304.
15. Oto, M., Koguchi, K., and Yuasa, Y. (1997) Analysis of a polyadenine tract of the TGF-b type II receptor gene in colorectal cancers by non-gel sieving capillary electrophoresis. *Clin. Chem.* **43,** 759–763.
16. Kuypers, A. W. H. M., Willems, P. M. W., van der Schans, M. J., Linssen, P. C. M., Wessels, H. M. C., de Bruijn, and C. H. M. M. (1993) Detection of point mutations in DNA using capillary electrophoresis. *J. Chromatogr.* **621,** 149–156.
17. Blin, N. and Stafford, D. M. (1976) A general method for the isolation of high molecular weight DNA from eukaryotes. *Nucleic Acids Res.* **3,** 2303–2308.

15

Detection of Microsatellite Instability in Cancers by Means of Nongel-Sieving Capillary Electrophoresis

Michiei Oto

1. Introduction

Microsatellite instability has been shown to be relevant to various human diseases, including fragile X syndrome *(1)* and Huntington's disease *(2)*. In several human cancers, it has been reported that an increase or decrease in the number of repeat units between lymphocyte and tumor DNA derived from the same patient was found *(3,4)*. Therefore, the analysis of microsatellite DNA has become necessary for the diagnosis of various diseases, especially hereditary and sporadic cancers *(3,4)*. In hereditary nonpolyposis colorectal cancer (HNPCC), abnormalities of microsatellite DNA, (CA)n repeats, were reported to be an effective marker of microsatellite instability *(5,6)*. The (CA)n repeats, which are related to HNPCC, are located on D2S123 marker DNA of the second chromosome. It is supposed that the alteration of (CA)n repeats might be associated with a defect in an early step of mismatch repair, and is controlled by mutator gene products such as hMSH2 *(11)*. For the analysis of microsatellite instability, the PCR technique has been applied *(5,8)*. DNA fragments containing (CA)n repeats can be amplified using specific primers *(8;* **Fig. 1**). The increase or decrease in the (CA) repeat number has been determined by evaluation of amplified DNA fragments by gel electrophoresis *(8)*.

However, for routine analysis of microsatellite DNA, the development of an automated analytical system is indispensable. As for pre-PCR steps, automated DNA extraction methods and robotic manipulation have been developed, and are commercially available. On the other hand, for post-PCR steps, gel electrophoresis, followed by ethidium bromide or silver staining, is generally per-

From: *Methods in Molecular Medicine, Vol 27: Clinical Applications of Capillary Electrophoresis*
Edited by: S. M. Palfrey © Humana Press Inc., Totowa, NJ

A

```
D2S123-U:   5'-GGATGCCTGCCTTTAACAGT-3'
D2S123-L:   5'-GACTTTCCACCTATGGGACT-3'
```

B

```
5'-[D2S123-U]-(CA)nCACACACACACACACA----------- 3'
3'------------(GT)nGTGTGTGTGTGTGTGT-[D2S123-L]-5'


5'-[D2S123-U]-(CA)nCACACACA----------- 3'
3'------------(GT)nGTGTGTGT-[D2S123-L]-5'
```

Fig. 1. Detection of (CA)n repeat "D2S123" alterations. (**A**) Primers used for amplification of (CA)n repeats. The sizes of the amplified DNA were 180–350 bp. These specific sequences are given at the side of (CA)n repetitive sequence. (**B**) Amplified DNA fragments containing different (CA)n repeat numbers. An increase or decrease in the (CA)n repeat number brings about a change in the size of the PCR product directly.

formed to separate and detect nanogram–picogram quantities of DNA. However, gel electrophoresis is not suitable for routine diagnosis, because it is difficult to automate.

In contrast to gel electrophoresis, capillary electrophoresis (CE) is a promising method for the high-speed and reproducible separation of DNA, and its use is feasible for automated DNA diagnosis (*9*). There are two types of CE, gel-sieving and nongel-sieving. Gel-sieving CE has some disadvantages in reproducibility and reusability, because of the difficulty of gel preparation. Nongel-sieving CE is a specialized form of capillary zone electrophoresis performed with a buffer containing polymer additives, such as methylcellulose (*10*). These additives change the rate of migration, depending on the molecular mass. A nongel-sieving capillary can be prepared by only introducing a buffer containing polymer additives into a capillary, and this technique gives reproducible results, and the capillary is reusable (*10,11*). In electrophoresis, DNA is monitored with a UV detector or a laser-induced fluorescence detector, and the acquisition of data, including their analyses, is computer-controlled, which facilitates automated DNA detection in multiple samples without a staining procedure.

This technique can allow the separation of single-stranded as well as double-stranded DNA in various size ranges within 30–40 min, under the optimal conditions; hence, the nongel-sieving CE technique is useful for the analysis of

DNA, including PCR products, and has been applied to various DNA diagnostic techniques *(11–14)*.

Extensive purification of amplified DNA by ultrafiltration or phenol extraction before CE has been performed to avoid peak broadening, and interference resulting from the presence of proteins and high concentrations of salts in the samples. An additional purification step is, however, a disadvantage, especially in the analysis of a large number of samples. The author investigated the conditions for the direct use of unpurified PCR products for CE analysis *(11)*.

In this chapter, the diagnosis of microsatellite instability using high-speed (CA)n repeats analysis is described, which combines modified rapid-cycle PCR and automated CE with the nongel-sieving technique, and the application to DNA diagnosis of colorectal cancer patients.

2. Materials

2.1. Reagents and Instrumentation for the PCR Technique

1. Cyclone Plus (Millipore, Bedford, MA) for the preparation of oligonucleotide primers.
2. Oligo Pak (Millipore) for the purification of oligonucleotide primers.
3. Sterilized distilled water for dissolving the primers.
4. Zymoreactor II incubator (ATTO, Tokyo, Japan) to carry out DNA amplification.
5. Deoxyribonucleotide triphosphates (Toyobo, Osaka, Japan).
6. Tris-HCl buffer: 10 mM Tris-HCl (pH 8.8), 50 mM KCl, 1.5 mM MgCl$_2$, and 0.1% gelatin.
7. Taq DNA polymerase (Toyobo).
8. Reaction tubes (Robbins Scientific, Sunnyvale, CA) suitable for the incubator.
9. Wax: AmpliWax™ (Perkin Elmer/Cetus, Foster City, CA).

2.2. Reagents and Instrumentation for Capillary Electrophoresis

1. CE was performed on a 50 μm (id) × 36 cm coated-capillary column (Bio-Rad, Hercules, CA).
2. 3X TBE buffer: 0.267 M Tris/borate, pH 8.3, 1 mM EDTA containing polymer additives (PCR Product Analysis Buffer™, Bio-Rad).
3. Laser fluorescence detector: BioFocus™ 3000 LIF^2 (Bio-Rad).
4. Washing solution: Capillary Washing Solution, Bio-Rad.
5. Standard DNA for calibrating repeated sequence: pBR322/*Ava*II/*Eco*RI (Bio-Rad). This contains 222–1746bp dsDNA.

3. Methods

3.1. Specimens and Genomic DNA Preparation

1. Tumor specimens from patients were obtained by means of surgical resection, according to the procedures of the Helsinki Declaration of 1975, as revised in 1983. High-mol-wt DNA was prepared by the method described previously *(15)*.
2. Lymphocyte DNA from the respective patients was also similarly prepared.

3.2. (CA)n Repeats Amplification Using the Modified Rapid-Cycle PCR Technique

A rapid-cycle PCR technique has been reported *(16,17)*, which needs specialized equipment. It was modified using a regular program incubator (**Fig. 2**).

1. Forward primer sequence, 5'-GGATGCCTG CCTTTAACAGT-3'; and Reverse primer sequence, 5'-GACTTTCCACCTATGG GACT-3' (**Fig. 1**).
2. In a thin-walled tube, mix oligonucleotide primers (160 n*M*), 20 µ*M* dNTP, 0.5 units of Taq DNA polymerase, 10 m*M* Tris-HCl, pH 8.8, to give a total volume of 50 µL (*see* **Notes 2** and **3**).
3. Add a piece of AmpliWax, and warm the tube to 70°C for 1 min to melt the wax, then keep at room temperature to solidify the wax.
4. To obtain constant results, and for application to routine assaying, store the preparations in multiple tubes at 4°C until the PCR experiments.
5. Add 1–2 µL of the template DNA (100–500 ng) onto the wax in the stored tubes.
6. Carry out 40 cycles of reaction at 94°C for 10 s, 60°C for 20 s, and 72°C for 10 s in the program incubator (*see* **Note 1**).
7. Pierce the solid wax with the pipet tip, and take the amplified products out, and then subject them to CE analysis without further purification.

3.3. CE Analysis of PCR Products

1. Prior to sample migration, the washing solution containing a detergent, distilled water, and the sieving buffer containing polymer additives are filtered, and the samples are degassed by centrifugation.
2. As shown in **Fig. 3**, wash the capillary for 30 s with the washing solution, and then for 30 s with distilled water, before each run. After washing of the capillary, introduce the buffer containing polymer additives into the capillary by means of a pressurized injection at 80 psi for 120 s. These processes are programmed with software, and are automatically performed with a CE instrument.
3. Introduce the DNA sample by pressurized injection at 80 psi, 10 nL of the DNA solution are introduced into the capillary for analysis.
4. The electrophoresis is performed for 15 min at 222 V/cm at 30°C in TBE buffer (0.267 *M* Tris-borate, pH 8.3, 1 m*M* EDTA) containing polymer additives. Set the UV detector to 260 nm, with a range of 0.02 absorbance units (*see* **Notes 4–7**).

3.4. Data Analysis

Transfer stored raw data from braided file format (BFF) to a format compatible with the BioFocus Integrator on a computer. Perform postrun analysis of the data using the BioFocus Integrator software (version 6.0). The fragment sizes are determined with the software, using a linear regression method to establish a curve of best fit, generated with the standards in each series of analyses.

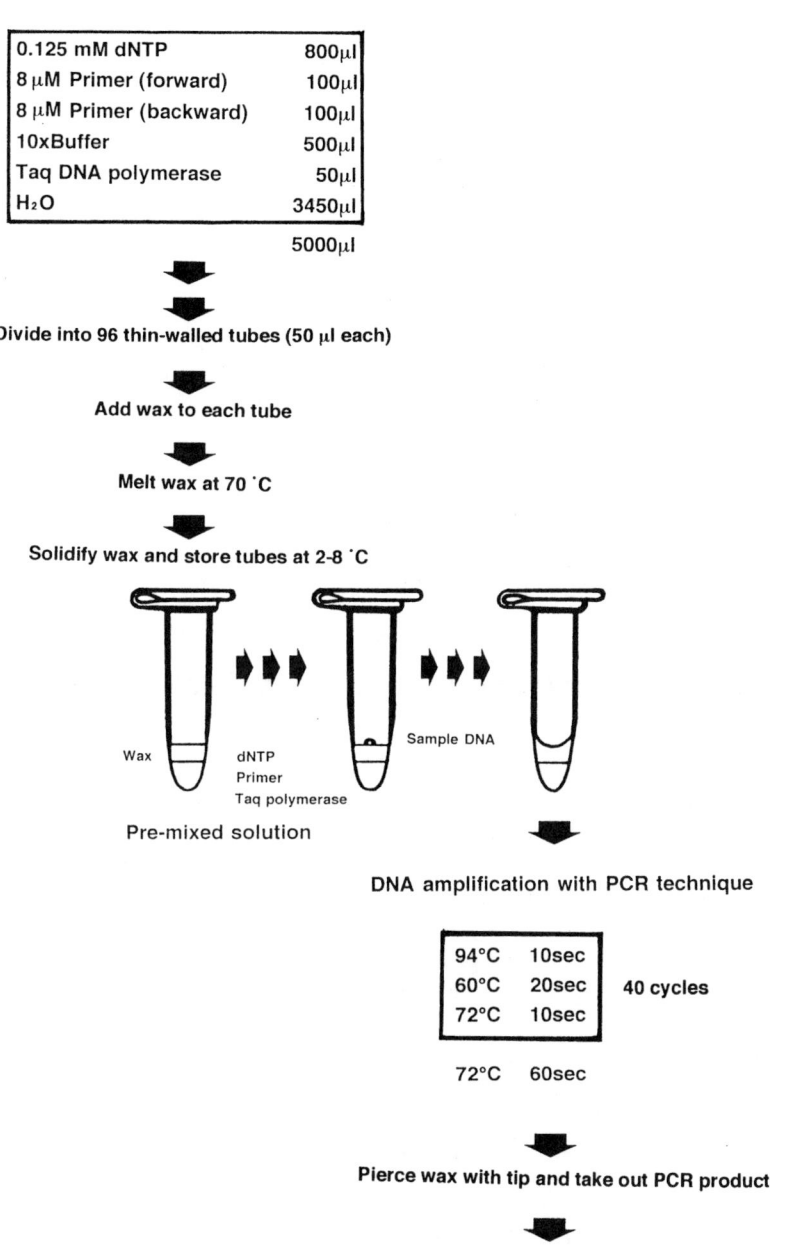

Fig. 2. Modified rapid cycle PCR technique. The 10X buffer comprised of 100 m*M* Tris/HCl, pH 8.8, containing 500 m*M* KCl, 15 m*M* MgCl, and 1% gelatin. The PCR conditions, including the annealing temperature, which were optimized for the amplification of D2S123, are shown.

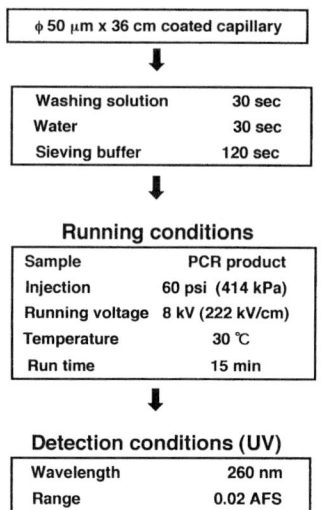

Fig. 3. (CA)n repeat analytical conditions using nongel CE.

3.5. Typical Results for Detection of Microsatellite Instabilities in Colorectal Cancers

1. The results of CE analyses of amplified D2S123 DNA, derived from various colorectal cancer patients, are shown in **Fig. 4**. Compare the electrophoretic profiles of the amplified products between tumor and normal DNA from the same patient. In case A, the mobility of tumor DNA is faster than that of normal DNA. In case B, the largest DNA fragment in normal cells has disappeared in the tumor. On the other hand, the mobilities of tumor and normal DNA are identical in case C.
2. These profiles of amplified products observed on CE are the same as, or better than, the data obtained on gel electrophoresis (data not shown). These data indicate that the microsatellite instability in cancers can be detected on CE in much the same way to as on gel electrophoresis, and that the time required for CE analysis is about 10 min, i.e., much shorter than that for gel electrophoresis. In case B, the disappearance of the largest DNA may indicate loss of heterozygosity (LOH). However, further studies are necessary to determine whether this phenomenon is genetic instability or LOH.

3.6. Confirmation of (CA)n Repeat Differences Using LIF-CE

To confirm the differences in the profiles of the two PCR products (e.g., derived from normal and tumor tissues), amplify each DNA with primers labeled with FITC and Texas Red, respectively. The LIF-CE conditions are shown in **Fig. 5**. In case A, the electropherogram clearly shows the differences in mobility (**Fig. 6**).

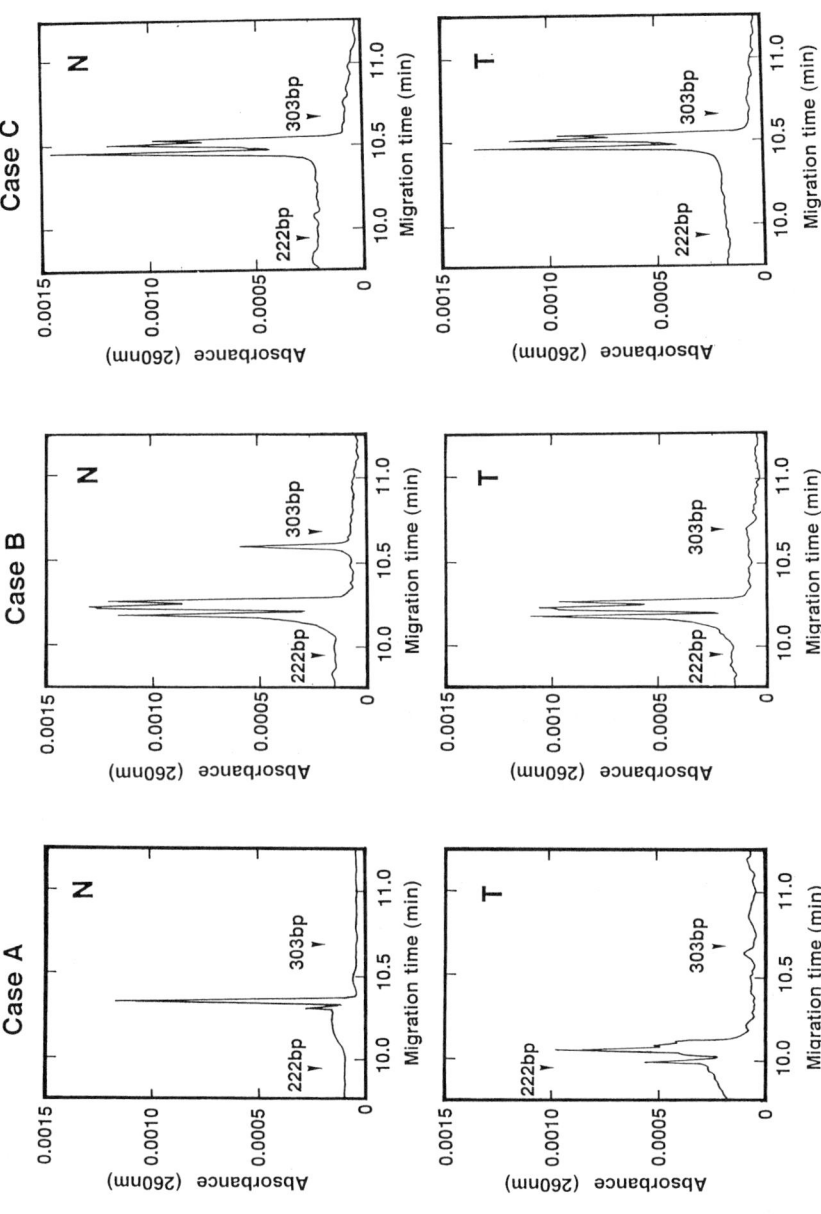

Fig. 4. Non-gel sieving CE analysis of (CA)n repeats amplified with the modified rapid-cycle PCR from various specimens. electrophoretic conditions were the same as given in **Fig. 3**. Amplified DNA was derived from normal (N) and tumor (T) cells f the three representative colorectal cancer patients. The sizes of pBR 322/*Ava*II/*Eco*RI markers are indicated.

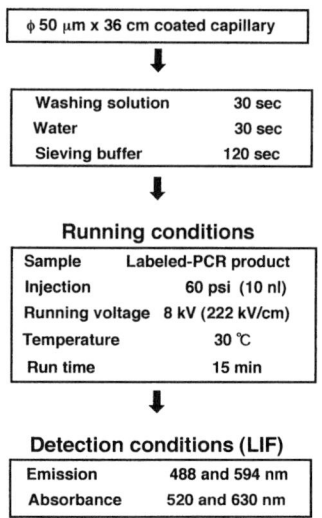

Fig. 5. (CA)n repeat analytical conditions using nongel CE with a LIF detector.

3.7. Advantages of This Method

The microsatellite instability analysis method described here.has several advantages. For analysis with CE, nongel-sieving CE is preferable, in both reusability and reproduciblity, to gel sieving CE. PCR products can be subjected to nongel-sieving CE without further purification, and analyzed in about 10 min, with the same or better resolution as that on gel electrophoresis. With this method, postseparation detection, such as ethidium bromide staining or silver staining, is not needed.

In HNPCC, and in some sporadic colorectal cancers with microsatellite instability, one or two deletions of a polyadenine tract, (A)10 repeats, of the transforming growth factor-β type II receptor gene, were recently reported *(14)*. One base deletion of a (A)10 repeat was also detected on nongel-sieving CE *(14)*.

Hence, the nongel-sieving CE technique is preferable for the analysis of repeated sequences, because of higher resolution and easy manipulation.

4. Notes

1. By gel electrophoretic analysis, the amplified (CA)n repeats obtained with this method were determined to be of the same quality as those obtained with the regular PCR method (data not shown). However, if longer-sized fragments (more than 500 bp) are amplified, the elongation time is recommended to be longer (72°C, 15–25 s).
2. With the pre-mixed solution, which was stored for 1, 2, or 4 mo at 4°C, DNA was amplified as in the case of the premixed one immediately after preparation (data not shown).

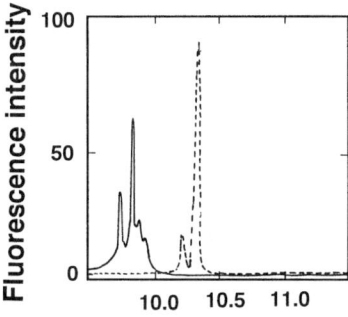

Migration time (min)

Fig. 6 Typical electropherogram (LIF-CE). The differences in the profiles of normal and tumor DNA with the double-injection technique. The specimens are from case A. The electropherogram of DNA from normal and tumor tissues are shown by the dotted line and the solid line, respectively.

3. Prior to CE analysis, the purification of the PCR products by ultrafiltration or phenol extraction was performed to avoid peak broadening and interference by the high concentrations of salts in samples. However, this step is disadvantageous for the analysis of a large number of samples. For application to routine analysis, the author examined the use of unpurified PCR products for CE analysis (*11*). When the author applied the unpurified (CA)n repeats amplified under ordinary conditions, extra peaks were found in front of the target peaks (**Fig. 7A**). When 200 µ*M* dNTP and 160 n*M* primers were applied separately to study the extra peaks, 200 µ*M* dNTP gave the same peaks as the extra ones (**Fig. 5B**), but the 160 n*M* primers did not (data not shown). Therefore, PCR was carried out with 20 µ*M*, instead of 200 µ*M* dNTP. The PCR products gave the same target peaks, but not the extra peaks (**Fig. 7C**). Excessive peak broadening, caused by the differences in the salt concentrations between the salt-containing PCR products and the running buffer, was not observed (data not shown). These data indicate that the PCR products with 20 µ*M* dNTP can be subjected to CE without further purification. Therefore, the optimum concentration of dNTP was shown to be 20 µ*M* (final concentration) (*see* **Subheading 3.**) for the modified rapid cycle PCR technique.
4. In order to determine the reproducibility of this technique, three independent (CA)n repeats were amplified, and CE was separately performed for each product 10× under the optimum conditions. The CV of the migration time was shown to be within 0.7% (data not shown), indicating good reproducibility. To obtain better reproducibility, one should use a larger volume of running buffer (e.g., 0.5 mL/vial) to make the changes in salt concentrations during a run minimum.
5. If one reuses a capillary more than 100×, an acryloyaminothoxyethanol (AAEE)-coated capillary (Bio-Rad) is recommended, because of its long life. The use of this capillary brought good reusability (*18*).

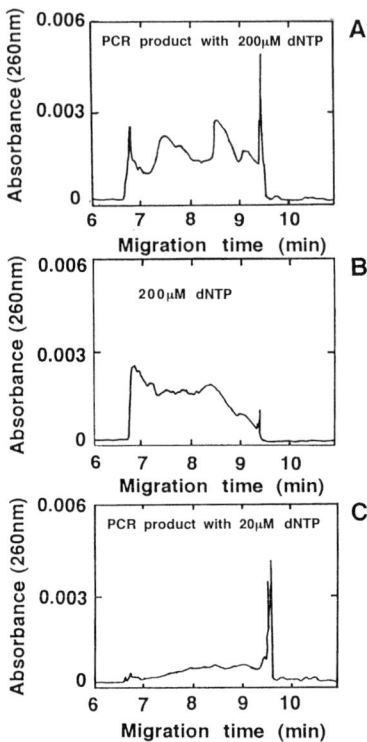

Fig. 7. Resolution of (CA)n repeats amplified with different dNTP concentrations on nongel-sieving CE. (**A**) PCR products amplified with 200 m*M* dNTP. (**B**) 200 m*M* dNTP only. (**C**) PCR products amplified with 20 m*M* dNTP.

6. Because a buffer containing polymer additives is highly viscous, the buffer containing polymer additives sometimes becomes attached to the surface of a capillary, depending on the concentration. To prevent carryover of the viscous buffer into the sample, it is desireable to dip the capillary into distilled water 3×.

7. It is clear that the DNA fragment sizes to be separated should depend on running conditions, including voltage, temperature, and polymer concentration (**Fig. 8**). The sizes of amplified microsatellite DNAs vary. For example, the length of (CA)n repeats, "D2S123", fragments is 180–350 bp(8), but other markers have different ones (1,2). Therefore, the running conditions should be optimized for each amplified DNA.

In order to obtain higher resolution, standard DNA (pBR322/*Ava*II/*Eco*RI) was electrophoresed at various voltages, temperatures, and concentrations of polymer additives. The higher the voltage was, the shorter the running time. However, the resolution was worse at a higher voltage (**Fig. 8A**). To analyze

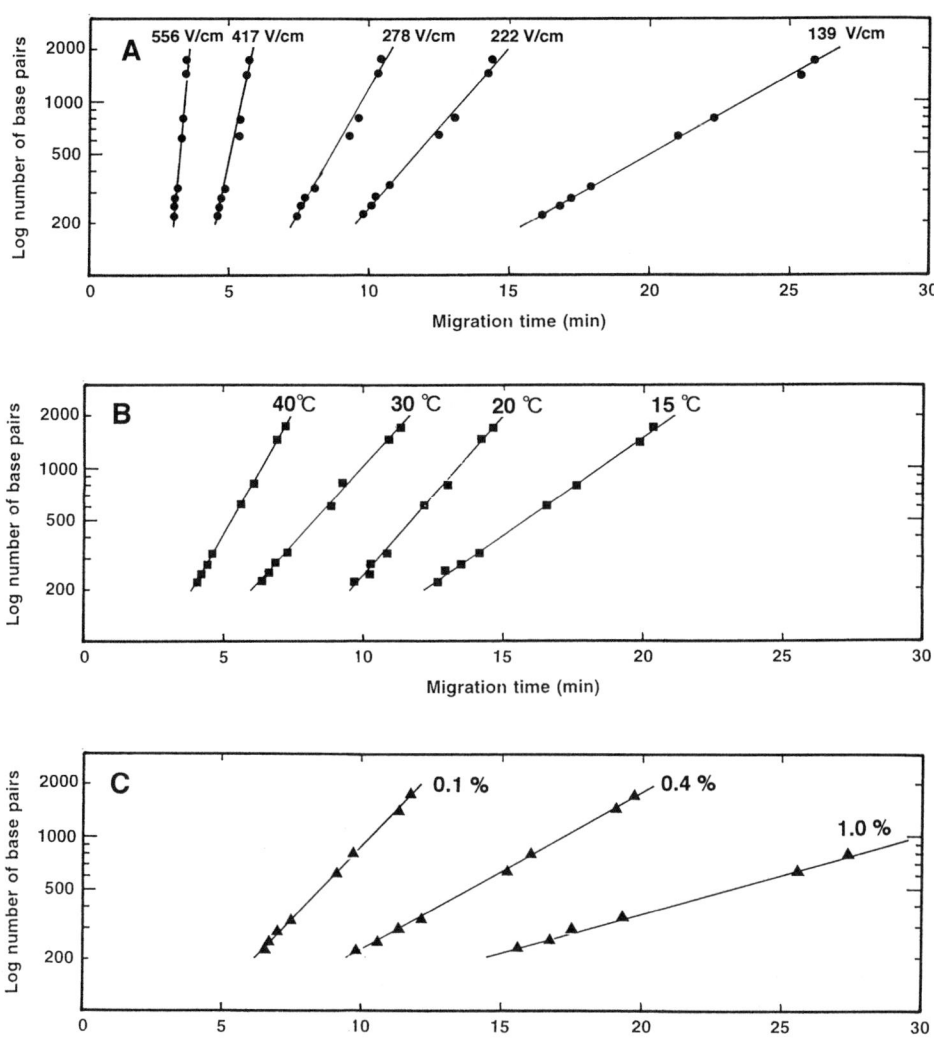

Fig. 8. Influence of various factors on electrophoresis. The standard DNA (pBR322/*Ava*II/*Eco*RI) was electrophoresed under various conditions. (**A**) Effects of voltage on the analysis time and resolution. The electrophoretic voltages were 139 V/cm, 222 V/cm, 278 V/cm, 417 V/cm, and 556 V/cm. The running temperature was 30°C. (**B**) Effects of temperature on the resolution. The electrophoretic temperatures were 15°, 20°, 30°, and 40°C. The running vaoltage was 222 V/cm. Effects of the concentrations of polymer additives (**C**). Hydroxy-propyl-methyl cellulose, as polymer additives, was used. The running voltage and temperature were 222 V/cm and 30°C, respectively.

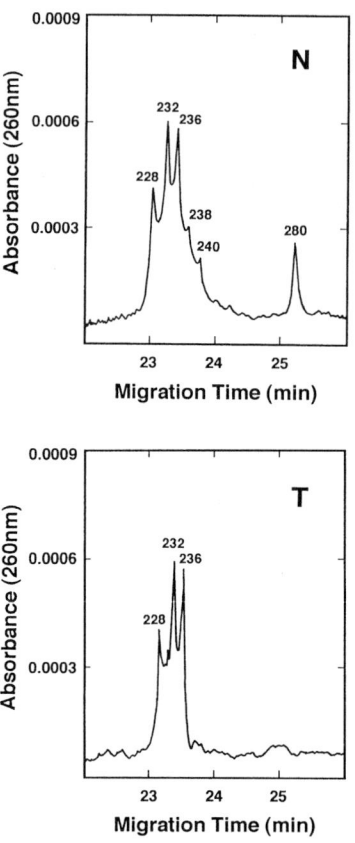

Fig. 9. Precise study of the case B specimen. Amplified DNA was derived from normal (N) and tumor (T) cells from case B. The PCR products were electrophoresed at 139 V/cm and 20°C for 30 min, using a buffer (0.267 *M* Tris/borate, pH 8.3) containing 1% HPMC. The DNA fragment sizes were determined with software, using a linear regression method to establish a curve of best fit generated for the standard DNA (pBR322/*Ava*II/*Eco*RI).

200–300 bp DNA of (CA)n repeats rapidly, sufficient resolution was obtained at 222 V/cm in 10 min. Better resolution was acquired at 139 V/cm, but it needed too long an analysis time. To determine the optimal running temperature at 222V/cm, standard DNA was run at 20–40°C. The best separation was obtained at 30°C (**Fig. 8B**). Figure 8C shows an electropherogram in which standard DNA was run using buffer containing various concentrations of hydroxypropylmethylcellulose (HPMC) as polymer additives. It is clear that higher resolution will be obtained, depending on the concentrations of polymer additives. The suitable temperature for both the capillary and the carousel will be 30°C, according to the author's experience, because the viscosity of various polymers may be stable at 30°C. The voltage and polymer concentrations can change the resolution dramatically. If a commercially available buffer containing polymer additives (e.g., PCR Analysis Buffer; Bio-Rad) is used, a lower voltage (e.g., 139 V/cm) and a longer capillary will bring about higher resolution. In **Fig. 9**, the precise analysis of amplified DNA from case B (**Fig. 4**) is shown. Differences of two (CA) units (4 bp) have been determined using 1% HPMC polymer under optimum electrophoretic conditions.

References

1. Pieretti, M., Zhang F., Fu, Y.-H., Warren, S. T., Oostra, B. A., Caskey, C. T., and Nelson, D. L. (1991) Absence of expression of the FMR-1 gene in fragile X syndrome. *Cell* **66,** 817–822.
2. The Huntington's disease collaborative research group. (1993) Novel gene containing a trinucleotide repeat that is expanded and unstable on Huntington's disease chromosome. *Cell* **72,** 971–983.
3. Han, H.-J., Yanagisawa, A., Kato, Y., Park, J.-G., and Nakamura, Y. (1994) Genetic instability in pancreatic cancer and poorly differentiated type of gastric cancer. *Cancer Res.* **53,** 5087–5089.
4. Wooster, R., Cleton-Jansen, A. M., Collins, N., Mangion, J., Cornelis, R. S., Cooper, C. S., et al. (1994) Instability of short tandem repeats (microsatellites) in human cancer. *Nature Genet.* **6,** 152–156.
5. Aaltonen, L. A., Peltomaki, P., Leach, F. S., Sistonen, P., Pylkkanen, L, Mecklin, J.-P., et al. (1993) Clues to the pathogenesis of famillial colorectal cancer. *Science* **260,** 812–816.
6. Peltomaki, P., Lothe, R. A., Aaltonen, L. A., Pylkkanen, L., Nystrom-Lahti, M., Seruca, R., et al. (1993) Microsatellite instability is associated with tumors that characterize the hereditary non-polyposis colorectal carcinoma syndrome. *Cancer Res.* **53,** 5853–5855.
7. Fishel, R., Lescoe, M. K., Rao, M. R. S., Copeland, N. G., Jenkins, N. A., Garber, J., Kane, M., and Kolodner, R. (1993) The human mutator gene homologue MSH2 and its association with hereditary nonpolyposis colon cancer. *Cell* **75,** 1027–1038.
8. Wu, C., Akiyama, Y., Imai, K., Miyake, S., Nagasaki, H., Oto, M., et al. (1994) DNA alterations in cells from hereditary non-polyposis colorectal cancer patients. *Oncogene* **9,** 991–994.
9. Martin, F., Vairelles, D., and Henrion, B. (1993) Automated ribosomal DNA fingerprinting by capillary electrophoresis of PCR products. *Anal. Biochem.* **63,** 182–189.
10. Zhu, M., Hansen, D. L., Burd, S., and Gannon, F. (1989) Factors affecting freezone electrophoresis and isoelectric focusing in capillary electrophoresis. *J. Chromatogr.* **480,** 311–319.
11. Oto, M., Suehiro, T., Akiyama, Y., and Yuasa, M. (1995) Microsatellite instability in cancer identified by non-gel sieving capillary electrophoresis. *Clin. Chem.* **41,** 482–483.

12. Oto, M., Suehiro, T., and Yuasa, Y. (1995) Identification of mutated p53 in cancer by non-gel-sieving capillary electrophoretic SSCP analysis. *Clin. Chem.* **41,** 1787–1788.
13. Oto, M., Suehiro, T., and Yuasa, Y. (1995) DNA hybridization analysis of PCR products by non-gel sieving capillary electrophoresis. *PCR Methods. Applic.* **4,** 303–304.
14. Oto, M., Koguchi, K., and Yuasa, Y. (1997) Analysis of a polyadenine tract of the TGF-b type II receptor gene in colorectal cancers by non-gel sieving capillary electrophoresis. *Clin. Chem.* **43,** 759–763.
15. Blin N. and Stafford, D. M. (1976) A general method for isolation of high molecular weight DNA from eukaryotes. *Nucleic Acids Res.* **3,** 2303–2308.
16. Wittwer, C. T., Fillmore, G. C., and Hillyard, D. R. (1989) Automated polymerase chain reaction in capillary tubes with hot air. *Nucleic Acids Res.* **17,** 4353–4357.
17. Wittwer, C. T., Fillmore, G. C., and Garling, D. J. (1990) Minimizing the time required for DNA amplification by efficient heat transfer to small samples. *Anal. Biochem.* **186,** 328–331.
18. Talmadge, K. W., Zhu, M., Olech, L., and Siebert, C. (1996) Oligonucleotide analysis by capillary polymer sieving electrophoresis using acryloylamino-ethoxyethanol-coated capillaries. *J. Chromatogr.* **744,** 347–354.

16

Serum Lamotrigine Analysis

Zak K. Shihabi

1. Introduction

Antiepileptic drugs vary greatly in chemical structure. Analysis of these compounds by capillary electrophoresis (CE) requires different conditions for each one, or for each group. Acidic drugs can be analyzed easily by capillary zone electrophoresis in borate buffers *(1)*; the neutral and the mixtures can be better analyzed by micellar electrokinetic chromatography *(2–4)*. Chapter 17 describes how phenobarbital and a few other antiepileptic drugs could be analyzed by CE.

Basic drugs, in general, are more difficult to analyze by CE and by chromatographic methods, because they bind to the surface of the packing material or the capillaries. An example of a basic antiepileptic drug used here is lamotrigine (Lamictal, Burroughs Wellcome, Research Triangle, NC). This is a new antiepileptic drug (3,5 diamino-6-dichlorophenyl-1,2,4-triazine) with a half-life of about 24 h. It is metabolized by glucuronide conjugation, with a peak level occurring at 2 h after an oral dose *(5–7)*. The therapeutic level of lamotrigine is not well established but patients who respond to this drug have levels of about 1–4 mg/L; however, most of the author's patients have a mean of 3.7 mg/L ±2.8 *(8)*.

HPLC methods for lamotrigine analysis require large amounts of organic solvents for elution, which are expensive and environmentally hazardous. Many of these methods also require extraction for sample cleanup before injection on the column. On the other hand, the analysis of lamotrigine by CE is simple, rapid, and avoids the (HPLC) problems *(8)*.

2. Materials

1. Internal standard: 40 mg/L Tyramine (Sigma, St. Louis, MO) in acetonitrile.

From: *Methods in Molecular Medicine, Vol 27: Clinical Applications of Capillary Electrophoresis*
Edited by: S. M. Palfrey © Humana Press Inc., Totowa, NJ

2. Electrophoresis buffer: 130 m*M* sodium acetate; adjust to pH 4.8 with 1 *M* HCl.
3. Stock standard: 10 mg lamotrigine is dissolved in 100 mL methanol (keep refrigerated).
4. Working standard: Dilute 1 mL of the stock standard in 9 mL serum free of this drug; store frozen (*see* **Note 1**).
5. Capillary: Untreated fused silica 42 cm × 50 μm (id).
6. CE instrument: A Model 2000 capillary electrophoresis instrument (Beckman, Fullerton, CA) is set at 10 kV. Running temperature 24°C and wavelength 214 nm.

3. Methods

1. To 0.5-mL centrifuge tubes, add 100 μL serum (standard or control) and 200 μL acetonitrile, containing 40 mg/L tyramine as an internal standard (*see* **Note 2**).
2. Vortex-mix for 15 s, and centrifuge the mixture for 30 s at 14,000*g*.
3. Transfer the supernatant to a sample vial, and add 200 μL acetic acid, 0.9 *M*; mix, and inject into the capillary (*see* **Note 3**).
4. Condition a new capillary by rinsing with 0.2 *M* NaOH for 20 min, then with water for 2 min, followed by separation buffer for 5 min.
5. Introduce the sample into the capillary by low pressure (0.5 psi) injection for 25 s (*see* **Note 4**).
6. After each sample, the capillary is washed for 1 min with 0.2 *M* NaOH, followed by electrophoresis buffer for 1 min.
7. **Figure 1** illustrates the detection of lamotrigine in serum (*see* **Note 5**).
8. The limit of detection is 1 mg/L (*see* **Note 6**).

4. Notes

1. Because of matrix effects and difference in recovery, the standards should be added to a serum sample free from this drug. Controls can be prepared as the working standards, using a different serum pool and different dilutions.
2. The internal standard is used to correct (or check) for the migration time, but it is not useful for correction for peak area or height.
3. Because, the pH of the supernatant is ~7.4 after serum precipitation, it requires acidification with the diluted acetic acid, in order to drop it below the pK of lamotrigine.
4. Ignore the first three injections after starting the instrument, because they have slow migration times.
5. Serum contains an unknown compound (X in **Fig. 1**), which migrates after, but close to, lamotrigine.
6. Newer models of CE instruments have about 4× less baseline noise than the author's early model (**Fig. 1**). Values below 1 mg/L are difficult to analyze by this instrument. To improve the sensitivity, several options can be used: extraction, use of 75 μm (id) capillary, a more optimum wavelength, and the use of a capillary with a Z or bubble cell.

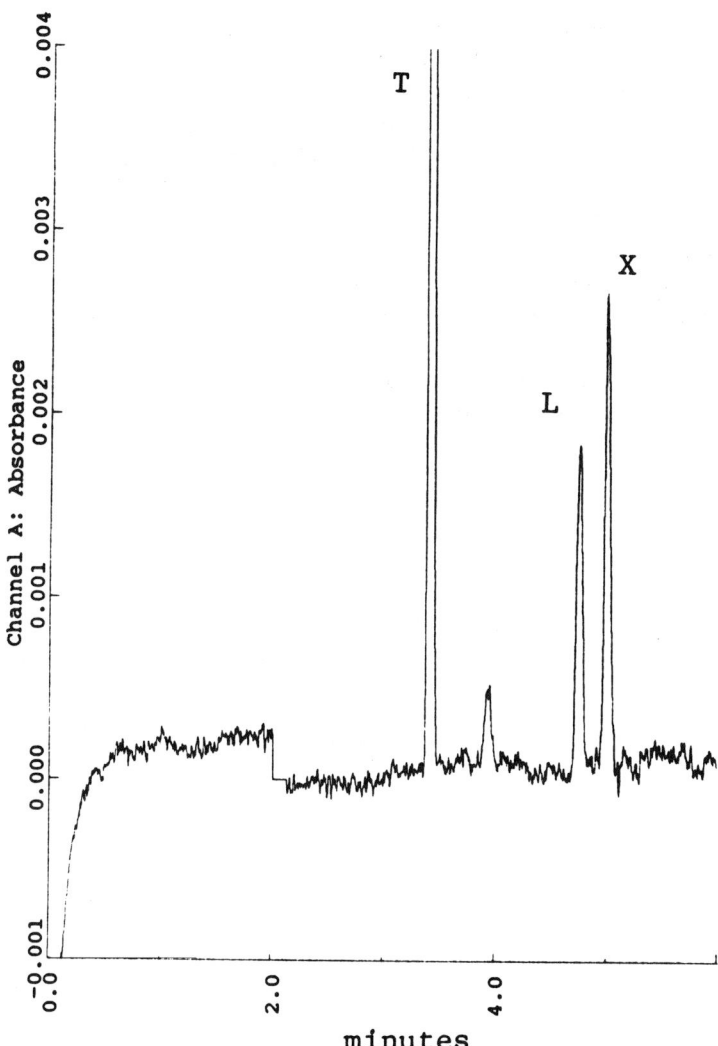

Fig. 1. Lamotrigine (8.4 mg/L) from a patient serum (T, tyramine; L, lamotrigine; X, unknown endogenous peak).

References

1. Shihabi, Z. K. (1993) Serum pentobarbital determination by capillary electrophoresis. *J. Liq. Chromatogr.* **16,** 2059–2068.
2. Shihabi, Z. K. and Oles, K. S. (1994)Felbamate measured I serum by two methods: HPLC and capillary electrophoresis. *Clin. Chem.* **40,** 1904–1908.
3. Thormann, W., Meier, P., Marcolli, C., and Binder, F. (1991) Analysis of barbiturates in human serum and urine by high performance capillary electrophoresis-

micellar electrokinetic capillary chromatography with on-column multi-wavelength detection. *J. Chromatogr.* **545,** 445–460.

4. Shihabi, Z. K. (1997) Therapeutic drug monitoring by capillary electrophoresis, in *Handbook of Capillary Electrophoresis Applications* (Shintani, H. and Polonsky, J., eds.), Chapman and Hall, London, pp. 386–408.

5. Messenheimer, J. A. (1995) Lamotrigine. *Epilepsia* **36,** S87–S94.

6. Pellock, J. M. (1994) Clinical efficacy of lamotrigine as antiepileptic drug. *Neurology* **44,** S29–S35.

7. Devinsky, O., Vazquez, B., and Luciano, D. (1991) New antiepileptic drugs for children: felbamate, gabapentin, lamotrigine and vegabatrin. *J. Child Neurol.* **9S,** S33–S45.

8. Shihabi, Z. K. and Oles, K. S. (1996) Serum lamotrigine analysis by capillary electrophoresis. *J. Chromatogr. B.* **683,** 119–123.

17

Acetonitrile Stacking

Serum Phenobarbital as an Example

Zak K. Shihabi

1. Introduction

Capillary electrophoresis (CE) has several advantages, such as high resolution, speed, and low cost of operation. However, it suffers from two major drawbacks: poor detection limits and matrix effects. Several approaches have been used to overcome these two problems. Here, acetonitrile stacking (AS) is presented as a simple, easy, and practical approach that can solve these two problems for many analytes, with minimum sample preparation. This laboratory successfully used AS for the analysis of several compounds by CE *(1)*.

1.1. Matrix Effects

Biological fluids such as serum contain high concentrations of proteins and salts. Resolution, theoretical plate numbers, precision, and quantification in CE are greatly affected by the composition of the sample especially its content of salts and proteins *(1,2)*. The salts cause band-broadening, because of the low field strength; proteins bind to the capillary walls and produce secondary interactions that greatly affect reproducibility. These effects become quite noticeable when the sample size increases to over 1% of the capillary volume *(1)*.

1.2. Stacking

Sample concentration on the capillary is called stacking. It is a very simple method to enhance detection, and can be produced in CE by several mechanisms *(1)*. In most instances, the analytes are induced to move rapidly, and stack as a sharp band as soon as the voltage is turned on; however, as the band enters the separation buffer, the different compounds separate into distinct bands.

From: *Methods in Molecular Medicine, Vol 27: Clinical Applications of Capillary Electrophoresis*
Edited by: S. M. Palfrey © Humana Press Inc., Totowa, NJ

1.3. Acetonitrile Stacking (AS)

This special type of stacking is achieved by including acetonitrile in the sample (not the buffer) *(1,3–5)*. Acetonitrile is often used in high-performance liquid chromotography (HPLC) for removing proteins from serum samples, in order to prevent pressure buildup, and to extend column-life. In addition to the removal of proteins, acetonitrile in CE has several other important advantages: It causes sample concentration and stacking, leading to ~10-fold enhanced detection, thus enabling many compounds to be amenable for CE; it decreases the need for a capillary wash; it improves the (RSD) by decreasing protein binding to the capillary; and it allows an increase in the sample volume.

1.4. Conditions for Acetonitrile Stacking

Because the injection volume in AS is large (typically 10–20% of the capillary volume), many factors in the sample itself can affect the separation. For example, the amount of acetonitrile, the pH, the ionic concentration, and the type of ion all affect the stacking *(6)*. Because many factors, such as ionic strength, salt composition, pH, amount of acetonitrile, and voltage, can affect the solubility, ionization, and migration of the different compounds in the sample, stacking responses can be different for each compound *(6)*. The following are some of the factors that can affect acetonitrile stacking:

1. Ionic strength in the sample: AS stacking is improved by an increase in the ionic strength of the sample *(2,4)*, provided acetonitrile is included in the sample itself *(4)*. Thus, this type of stacking is well suited for samples with high concentrations of salts or proteins, as is the case for serum or food. It also eliminates proteins that can interfere in the analysis, but it is limited to small molecules. The sample does not have to be in the same separation buffer. A different pH or buffer from that used in the separation can be desirable, in order to change the resolution or the plate number for a particular peak or component *(6)*.
2. Separation buffers: Although they slow the migration, strong separation buffers, in general, yield better separation and favor stacking *(2,5)*. Thus, indirectly, strong buffers allow a larger volume of the sample to be loaded on the capillary. Vinther and Soeberg *(7)* have described a mathematical model for the stacking force, based on the ratio of the electrophoresis buffer concentration to that of the sample.
3. Capillary length: The peak height depends on the capillary length. Since a larger sample volume can be loaded on a longer capillary, a greater concentration can be obtained on such a capillary *(4)*. Also, because of its length, a longer capillary gives a higher plate number, with better resolution.

1.5. Mechanism

The exact mechanism the produces that stacking is not completely clear, but is related to two factors: the low conductivity of acetonitrile, and the low solubility of the inorganic vs organic ions in the acetonitrile. The inorganic ions move rapidly as a wide band because of their limited solubility in the sample zone and for preserving the continuity of the current, while organic anions move behind, concentrating in the acetonitrile. This is analogous to transient isotachophoresis, in which a leading ion moves in high concentration, followed by the analytes ions. In the acetonitrile, the Cl– (or other inorganic ions) present in high concentrations in biological fluids, moves rapidly, acting as the leading ion, followed by the organic anions (analytes), concentrating as a sharp band. Unlike isotachophoresis, there is no critical demand in AS on controlling the order of the ion mobility or pH. Thus, the technique is simpler to perform than transient isotachophoresis. Another simple way to view AS stacking is that one edge of the sample moves at a different speed in relation to the other in order for the sample to concentrate. The sample edge in the inorganic ion region moves slowly (because of the low electrical field strength); that in the acetonitrile moves rapidly (because of the high electric field in that region), causing the sample to migrate as a sharp band.

1.6. Compounds Suitable for Acetonitrile Stacking

Because the exact mechanism is not well understood, it is difficult to predict the types of compounds amenable to, and conditions required for, stacking. In general, the author finds weakly anionic small molecules (e.g., the antiepileptic drugs) stack by this method. Several endogenous and exogenous compounds in the serum *(9–14)*, many enzymatic reactions *(15)*, and many peptides *(5)*, including insulin, stack well *(16)*. Some inorganic anions, such as bromide and nitrate *(8)*, also stack by this method. Recently, the author found that a high concentration (~0.5 *M*) of zwitterions, added to buffers, favors the stacking of many small cationic molecules, such as procainamide and quinidine(19).

2. Materials

1. Capillaries: Untreated fused-silica capillaries 40 cm × 50 μm (id) from Polymicro Technologies (Phoenix, AZ).
2. Instrument: A Model 2000CE (Beckman, Fullerton, CA) is set at 210 V/cm (*see* **Note 1**) and 214 nm.
3. Separation buffer: Borate 200 m*M*, pH 8.9 (*see* **Note 2**).
4. Stock drug solution: Phenobarbital 90 mg/L was dissolved in 50% methanol in water.
5. Internal standard: 20 mg/L iothalamic acid (Malinckrodt, St. Louis, MO) in acetonitrile.

3. Method

We will use the antiepileptic drug phenobarbital in serum as an example to illustrate how acetonitrile can be useful in CE.

3.1. Procedure

1. Add 100 μL serum to 200 μL acetonitrile containing 20 mg/L iothalamic acid (Malinckrodt) as an internal standard in a 0.5-mL microcentrifuge tube (*see* **Note 3**).
2. Vortex-mix the samples thoroughly for 30 s, and centrifuge for 30 s at 14000*g*.
3. Remove the supernatant, and transfer to the CE instrument (*see* **Notes 4** and **5**).
4. Fill the capillary with run buffer for 1 min.
5. Inject the sample by low pressure (0.5 psi) for the time required to fill about 10% of the capillary. **Subheading 3.2.** describes how to calculate the fill time (*see* **Note 6**).
6. Electrophorese for 12 min (*see* **Notes 7** and **8**).
7. After each run, rinse the capillary at high pressure for 2 min with 0.2 *M* NaOH, and with the electrophoresis buffer for 1 min.
8. **Figure 1** illustrates an example of the stacking of phenobarbital in serum (*see* **Note 9**).
9. Many other drugs can be analyzed by the same conditions, and migrate in the same time frame, such as pentobarbital *(17)*, phenytoin, theophylline, and iohexol.

3.2. Calculation of Injected Sample Size

1. The time to fill the capillary to the detector is determined experimentally by repeatedly injecting the sample with low pressure (0.5 psi) until the absorbance increases.
2. The % sample volume injected of the total capillary volume = (sample injection time/capillary fill time) × 100.

4. Notes

1. A low voltage favors stacking, but slows the analysis.
2. Prepare the buffer by dissolving the correct amount of boric acid, and adjusting the pH slowly to the desired pH with 2.5 *M* NaOH, without going above that value. If the pH goes over the desired pH, discard the buffer to avoid changes in the ionic strength. Ionic strength greatly affects the migration time in CE.
3. A lower ratio of acetonitrile to serum (e.g., 1:1) is not very effective in removing serum proteins *(3)*.
4. The samples should be analyzed immediately after preparation, unless they can be kept sealed on the instrument to avoid acetonitrile evaporation.
5. Not all the compounds stack with this procedure. Occasionally, the pH of the sample must be adjusted below the pK of the compound of interest to induce stacking *(6,18)*.
6. Usually, the sample size, under nonstacking conditions, is kept much below 1% of the capillary volume. Under AS, about one-third of the capillary volume can be filled with sample, with a corresponding increase in detection signal. With large sample injections, there is a slight decrease in plate number caused by the decrease in the capillary length. In practical analysis, numerous compounds

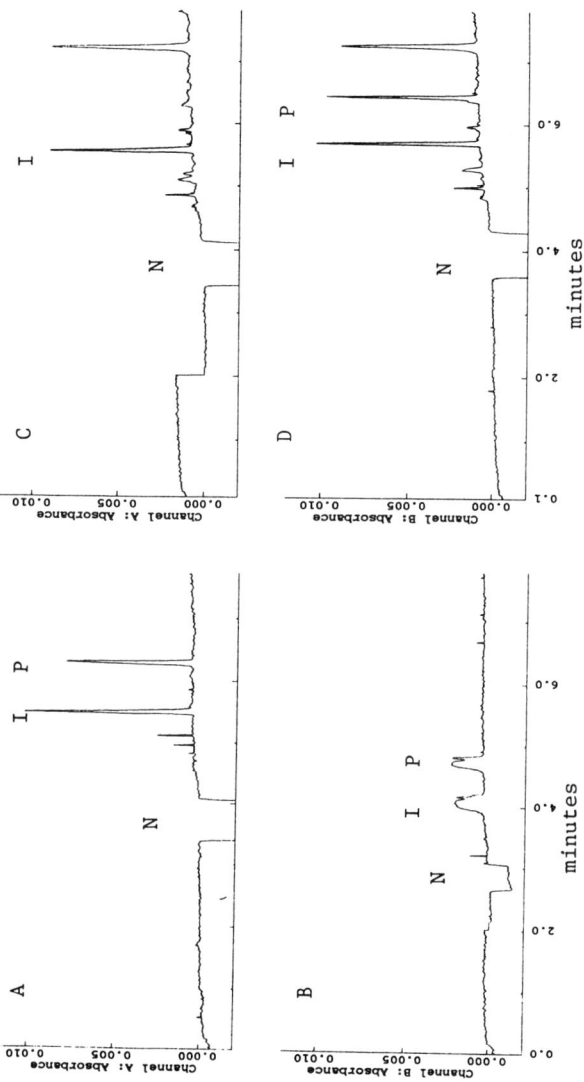

Fig. 1. This figure illustrates the stacking and analysis of phenobarbital. (**A**) The stock standard was diluted fivefold in 1% saline. A 100-µL aliquot was added to 200 µL acetonitrile containing 18 mg/L of iothalamic acid (internal standard), mixed, and injected, filling 10% of the capillary; (**B**) same as in (**A**), except water was substituted for acetonitrile; (**C**) 100 uL phenobarbital-free serum, mixed with 200 µL acetonitrile, vortex-mixed, centrifuged, and injected (filling 10% of the capillary); and (**D**) as in (**C**), but after spiking the serum with 18 mg/L of phenobarbital (within the therapeutic window).

Note: The injection is 10–20× greater than that used traditionally in CE. In the absence of acetonitrile, sample overloading is evident (wide peaks) in **Fig. 1B**. The presence of acetonitrile causes stacking (sharp peak), not just for the phenobarbital peak, but for most of the peaks present in the electropherogram. Note the peak width of phenobarbital, compared to that of the neutral compounds. N in the figure illustrates visually the degree of concentration (stacking).

present in complex samples, such as those found in serum or food, limit the sample loading to about 10% of the capillary volume. However, this loading is large enough to allow many drugs in serum to be detected at levels close to 1 mg/L.

7. A new capillary behaves different than an old capillary. This is caused by the strong adsorption of some compounds to the capillary. A new capillary is conditioned by rinsing before use with 0.2 *M* NaOH for 20 min, and with water for 2 min, then rinsing with the separation buffer for 5 min.

8. After a long shutdown, the first three injections are ignored, because they have slow migration times.

9. Occasionally, the migration time changes suddenly during the run. This is probably caused by the adherence of some components to the capillary walls. In this case, the capillary is washed as in **Note 2**.

References

1. Shihabi, Z. K. (1997) Effects of sample matrix on analysis, in *Handbook of Capillary Electrophoresis* (Landers J. P., ed.), 2nd ed., CRC, Boca Raton, FL, pp. 457–477.

2. Garcia, L. L. and Shihabi, Z. K. (1993) Sample matrix effects in capillary electrophoresis. I. Basic considerations. *J. Chromatogr. A.* **652,** 465–469.

3. Shihabi, Z. K. (1993) Sample matrix effects in capillary electrophoresis. II. Acetonitrile deproteinization. *J. Chromatogr. A.* **652,** 471–475.

4. Shihabi, Z. K. (1995) Sample stacking in acetonitrile-salt mixtures. *J. Cap. Electrophor.* **2,** 267–271.

5. Shihabi, Z. K. (1996) Peptide stacking by acetonitrile salt mixtures for capillary zone electrophoresis. *J. Chromatogr. A.* **744,** 231–240.

6. Friedberg, M. A., Hinsdale M., and Shihabi, Z. K. (1997) Effect of pH and ions in the sample on stacking in capillary electrophoresis. *J. Chromatogr. A.* **781,** 35–42.

7. Vinther, A. and Soeberg, H. (1991) Mathematical model describing dispersion in free solution capillary electrophoresis under stacking conditions. *J. Chromatogr.* **559,** 3–26.

8. Friedberg, M. A., Hinsdale, M. E., and Shihabi, Z. K. (1997) Analysis of nitrate in biological fluids by capillary electrophoresis. *J. Chromatogr. A.* **781,** 491–496.

9. Shihabi, Z. K. (1993) Serum pentobarbital determination by capillary electrophoresis. *J. Liq. Chromatogr.* **16,** 2059–2068.

10. Shihabi, Z. K. and Constantinescu, M. S. (1992) Iohexol in serum determined by capillary electrophoresis. *Clin. Chem.* **38,** 2117–2120.

11. Shihabi, Z. K., Hinsdale, M., and Bleyer, A. (1995) Xanthine analysis in biological fluids by capillary electrophoresis *J. Chromatogr. B.* **669,** 163–169.

12. Shihabi, Z. K. and Hinsdale, M. E. (1996) Analysis of ibuprofen in serum by capillary electrophoresis. *J. Chromatogr. B.* **683,** 115–118.

13. Friedberg, M. and Shihabi, Z. K. (1997) Ketoprofen analysis in serum by capillary electrophoresis. *J. Chromatogr. B.* **695,** 193–198.

14. Shihabi, Z. K., Rocco, M. V., and Hinsdale, M. E. (1995) Analysis of the contrast agent iopamidol by capillary electrophoresis. *J. Liq. Chromatogr.* **18,** 3825–3832.

15. Shihabi, Z. K. and Kute, T. (1996) Analysis of cathepsin D from breast tissues by capillary electrophoresis. *J. Chromatogr. B.* **683,** 125–131.
16. Friedberg, M. A. and Shihabi Z. K. (1998) Different methods for insulin stacking by capillary electrophoresis. *J. Chromatogr. A.* **807,** 129–133.
17. Shihabi, Z. K. (1997) Therapeutic drug monitoring by capillary electrophoresis, in *Handbook of Capillary Electrophoresis Applications* (Shintani, H. and Polonsky, J., eds.), Chapman and Hall, London, pp. 386–408.
18. Shihabi, Z. K. and Oles, K. S. (1996) Serum lamotrigine analysis by capillary electrophoresis. *J. Chromatogr. B.* **683,** 119–123.
19. Shihabi, Z. K.(1998) Stacking of weakly cationic compounds by acetonitrile for capillary electrophoresis. *J. Chromatogr. A.* **817,** 25-30.

18

Confirmation of the Presence of Drugs of Abuse in Urine

Stephen M. Palfrey

1. Introduction

Urine screenings for drugs of abuse are performed for many reasons, two of the most common being workplace/employment screening and monitoring in drug-dependence treatment centers. In both, the initial screen is commonly an immunoassay. Negative results require no further action. Positive tests require a second different method to both confirm and identify which drug is present. Some perfectly legal preparations crossreact with immunoassays. Employment screening has more stringent requirements than does monitoring of known drug abusers *(1)*, and gas chromatography–mass spectrometry (GC–MS) *(2,3)* is the method of choice. For the monitoring of known drug abusers attending drug-dependence centers, less sophisticated methods, such as thin layer chromatograghy(TLC) *(4)* and high-performance liquid chromatography (HPLC) *(5,6)*, are frequently used. Capillary electrophoresis (CE) methods have been described for the detection of drugs of abuse in pharmacological/illicit street samples *(7–9)*, and a few methods have been published using biological materials *(10–12)*. This chapter describes a simple sample preparation method, followed by CE for the confirmation and identification in urine of opiates, methadone (and its main metabolite EDDP), and amphetamines. Samples are prepared by a solid phase extraction procedure using columns packed with a co-polymeric material. This is a silica-based material to which is bound a mixture of hydrophobic and ionic groups. Acid and neutral drugs are bound through hydrophobic interactions only; basic drugs are bound by both hydrophobic and ionic interactions. Selectivity is increased by using an acidic wash buffer that ensures basic drugs are ionized, and are not removed by the methanol wash, which allows many interferences to be removed from the matrix before the isolate is eluted.

From: *Methods in Molecular Medicine, Vol 27: Clinical Applications of Capillary Electrophoresis*
Edited by: S. M. Palfrey © Humana Press Inc., Totowa, NJ

To elute the basic drugs, both the ionic and the hydrophobic interactions must be suppressed. This is achieved by using a relatively nonpolar organic eluant containing ammonium hydroxide, so that the eluant pH is alkaline (at least 2 pH units above the pKa of the drugs of interest). This pH neutralizes the charge on the basic isolates, allowing them to be easily removed from the matrix. After elution from the columns, they are evaporated to dryness and redissolved in water.

Opiates are separated in a fused silica capillary using a borate–phosphate buffer with the detergent sodium dodecyl sulphate as a micellar agent. The micelles are negatively charged, and tend to migrate toward the positive electrode. Electroosmotic flow transports the bulk solution toward the negative electrode, and is strong enough to overcome the electrophoretic migration of the micelles. Therefore, the anionic micelle travels toward the negative electrode at retarded velocity. The migration velocity of a drug, therefore, depends on the distribution coefficient between the micellar and aqueous phases.

Methadone and amphetamines are separated on the same capillary using free-solution electrophoresis in a borax buffer.

Confirmation of the identity of drugs is twofold. Comparison of peak migration times between unknowns and drug-free urines spiked with pure drugs is the first step. When there is a coincidence of migration time, the spectrum of the peak of the unknown can be compared with the spectrum of the corresponding peak in the spiked urine. The diode array is set to produce scans between 190 and 300 nm, as peaks are detected. Scans can be made on the rising slope, top, and falling slope of each peak. This is useful for checking the presence of more than one substance in the peak. Each of the different drug groups has distinctive spectra that makes confirmation of identity straightforward.

2. Materials

2.1. Sample Preparation

1. Solid phase extraction cartridges: Clean screen cartridges containing 200 mg solid phase, and with a 10-mL reservoir (Technicol, Stockport, UK).
2. Urine pH adjustment solution: 100 mM sodium phosphate, pH 6.0.
3. Column wash buffer: 100 mM acetate buffer, pH 4.2.
4. Column wash solvents: HPLC grade methanol, HPLC-grade water.
5. Elution solvent: Dichloromethane/propan-2-ol/conc ammonia (78/20/2,v/v/v). Prepare this solution fresh.
6. Stock drug standards 1 mg/mL methanolic solutions (Sigma,Poole, UK). Prepare working drug standards by diluting stock standards into a drug free urine. Store frozen at –20°C.
7. Vacuum elution manifold for solid-phase extraction.

2.2. Electrophoresis Buffers

1. Buffer for opiate separation: 80 mM boric acid, 30 mM sodium tetraborate (borax), 40 mM sodium dodecyl sulphate. Dissolve in about 700 mL HPLC-grade water, and adjust to pH 8.6. Add 200 mL acetonitrile, and make to 1 L with HPLC water. Stored at room temperature, tightly capped, this buffer is stable indefinitely.
2. Buffer for methadone–amphetamine separation: 65 mM borax: dissolve in 950 mL HPLC grade water, and adjust to pH 9.85 with 10 M NaOH. Make the solution to 1 L (*see* **Note 1**). Stored at room temperature tightly capped the buffer is stable indefinitely.

2.3. Electrophoresis Apparatus

1. The electropherograms and spectra reproduced in this chapter were created using a Beckman P/ACE MDQ (Beckman Instruments, Fullerton, CA) capillary electrophoresis system with a diode array detector.
2. Use a fused silica capillary (Supelco, Poole, UK), 50 μ id, 31-cm long (21 cm to detector window).

3. Methods

This cleanup procedure is a compromise designed to produce an aqueous sample that can be used for both the opiate and the amphetamine electrophoresis separations. The use of a pH 4.2 wash buffer is not optimal for amphetamines (*see* **Note 2**), but is still good enough to give a limit of detection of less than 0.1 mg/L. The limit of detection for opiates is 0.2 mg/L.

3.1. Sample Preparation

1. Pipet 2 mL standard/sample urine into a 10-mL plastic test tube, and add 4 mL of 100 mM phosphate buffer, pH 6.0 (the mixture should have a pH of 6.0–6.5) (*see* **Note 3**).
2. For each sample, condition a solid phase extraction (SPE) cartridge as follows: Wash with 3 mL of methanol, followed by 3 mL water, and finally with 2 mL 100 mM phosphate buffer, pH 6.0 (*see* **Note 4**).
3. Fill the reservoir with buffered sample, and adjust the vacuum to give a flow rate of 1–2 mL/min.
4. Wash each column with 2.0 mL water, followed by 2.0 mL 100 mM acetate buffer, pH 4.2, and finally with 3.0 mL methanol.
5. Adjust the vacuum to maximum, and dry the cartridges for 5 min.
6. Elute the samples from the cartridges with 3 mL elution solvent at 1–2 mL/min. Collect the eluate into glass tubes.
7. Evaporate to dryness at 70°C (*see* **Note 5**).
8. Redissolve the sample in 100 μL water, and transfer to a sample vial (*see* **Note 6**).

3.2. Electrophoresis

The buffer used in the opiate separation is similar to that used in a method to separate pharmacological samples of drugs *(11)*, but the ionic strength is significantly greater. This gives better separation of the different opiates, and alters the migration time of some minor interfering substances that are a potential source of interference with low (<1 mg/L) levels of opiates. Also, borate has been substituted for phosphate to reduce UV absorbance close to 200 nm. The amphetamine/methadone buffer is similar to one used by Molteni et al. *(12)*, but the ionic strength is greater and the pH higher, to optimize the separation of the various amphetamines.

3.2.1. Opiate Method

1. Fill the capillary with the borate buffer, using 20 psi pressure for 0.5 min. Load the sample for 3 s using 0.5 psi pressure.
2. Apply 12 kV (0.4 kv/cm), with the positive electrode at the inlet end of the capillary (*see* **Note 7**). Separation takes approx 7.5 minutes.
3. The diode array is set as follows: wavelength 214 nm (band width 20 nm), peak detect 214 ± 20 nm, scan range 190–300 nm.

3.2.2. Methadone–Amphetamine Method

1. Fill the capillary with the 65 m*M* borax buffer, using pressure for 0.5 min. Sample is loaded electrokinetically, using 2 kV for 5 s (*see* **Note 8**).
2. Apply 8 kv, with the positive electrode at the inlet end of the capillary. Separation takes approx 3.5 min.
3. The diode array is set as follows: wavelength 200 nm (band width 20 nm), peak detect 210 ± 20 nm, scan range 190–300 nm.

3.2.3. Capillary Maintenance

For neither method is it necessary to do anything but flush with electrophoresis buffer between samples (*see* **Note 9**). During long runs, ion depletion of the buffer can occur, and a change of buffer may become necessary. With the Beckman MDQ, this can be programmed into the method and multiple buffer vials preloaded. At the end of the run, flush the capillary with 0.2 *M* NaOH for 5 min, followed by water for 1 min.

3.3. Postanalytical

3.3.1. Opiates

The method described for the opiates detects morphine, monoacetyl-morphine, codeine, and hydrocodeine (*see* **Notes 2**, **10**, and **11**). Examples of electropherograms obtained from the urines of heroin, codeine, and hydrocodeine abusers are shown in **Figs. 1–3**. In addition to morphine (except when the amount present is small), heroin abusers always show smaller peaks of monoacetyl-morphine and codeine. Codeine and hydrocodeine abusers show peaks of nor-codeine and nor-hydrocodeine, respectively. The smallest amount detectable is 0.2 mg/L.

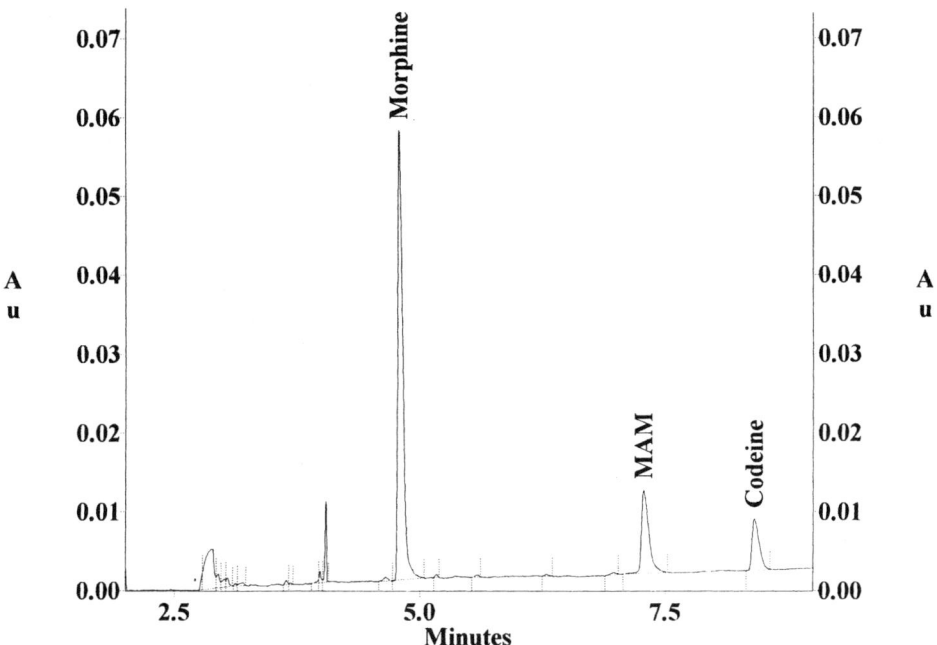

Fig. 1. Electropherogram of urine from a heroin abuser.

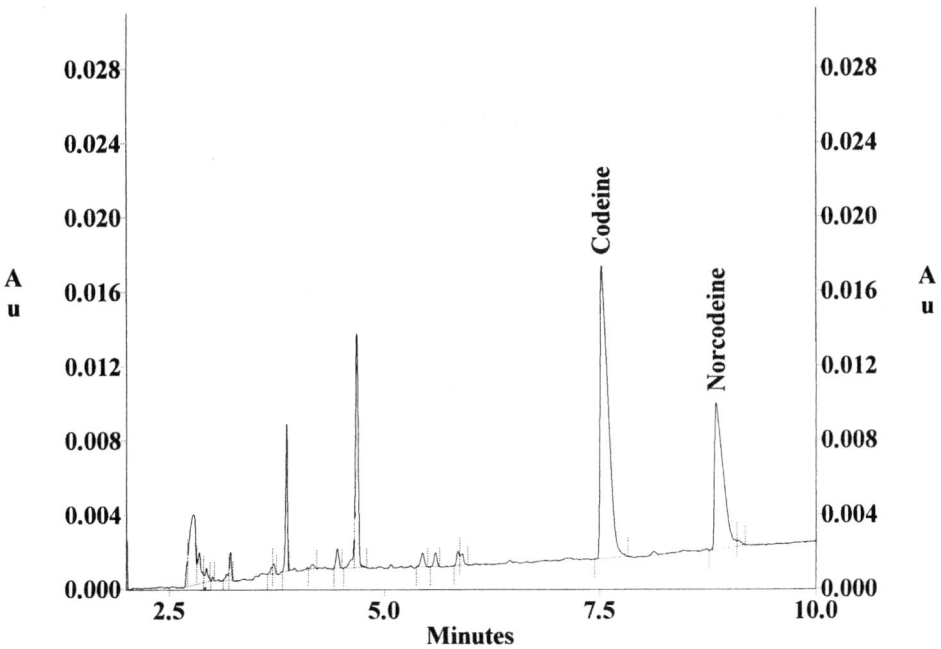

Fig. 2. Electropherogram of urine from a codeine abuser.

169

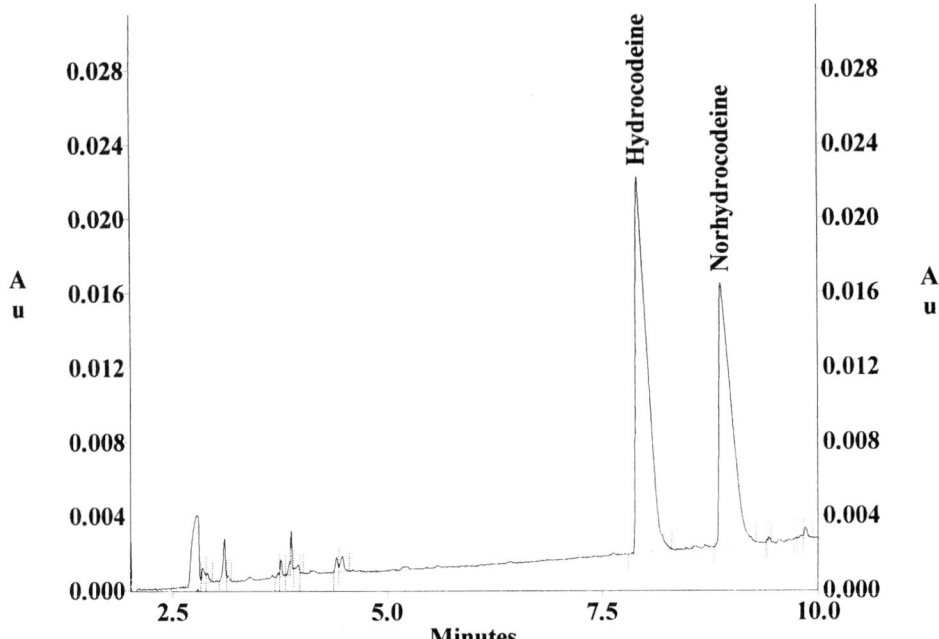

Fig. 3. Electropherogram of urine from a hydrocodeine abuser (extract diluted 20×).

3.3.2. Methadone and Amphetamines

1. In addition to methadone and its main metabolite, EDDP, this method also detects amphetamine and methamphetamine, plus the newer ecstasy-type amphetamines, MDMA, MDA, and MDE. Detection of EDDP in patients receiving methadone is important to confirm compliance. It is not unknown for a urine to be spiked with methadone syrup in an attempt to give a positive result. In this case, there will be no EDDP present.
2. **Figure 4** shows a drug free urine spiked with 0.5 mg/L of each drug detectable by this assay. **Figure 5** is an electropherogram from a patient on methadone, who is also taking amphetamine. **Figure 6** shows a urine from a patient who has taken methadone (*see* **Note 12**) and MDMA (Ecstasy). This assay can detect less than 0.1 mg/L of each drug (*see* **Note 8**).

3.3.3. Diode Array Scans

1. The diode array provides additional confirmation of drug identity, particularly when drugs have similar migration times. All of the drug groups detected by the methods described here have distinctive spectra. However, some drugs within a group have similar spectra.
2. The opiates, morphine and codeine, have very similar spectra, but fortunately have widely different migration times. Codeine and hydrocodeine have similar

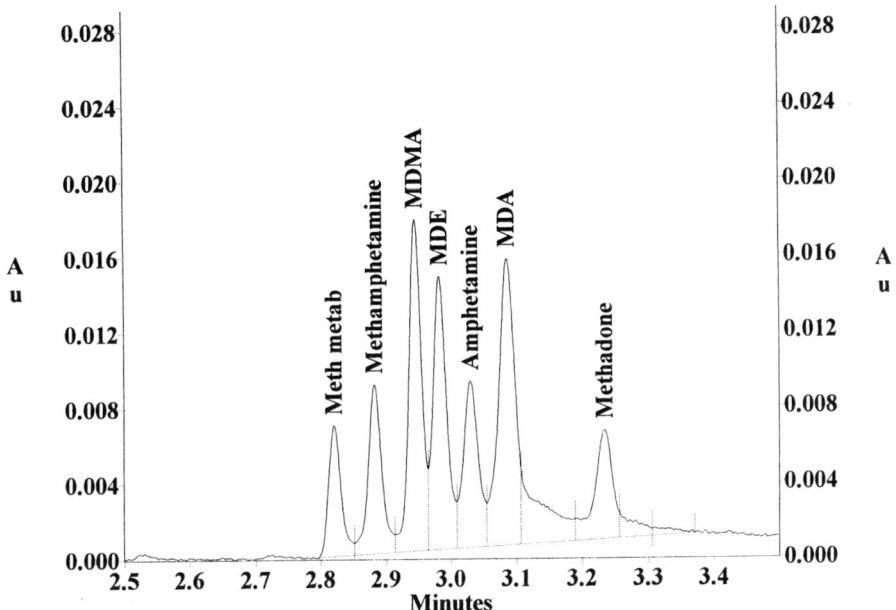

Fig. 4. Electropherogram of drug-free urine spiked with 0.5 mg/L of each drug shown.

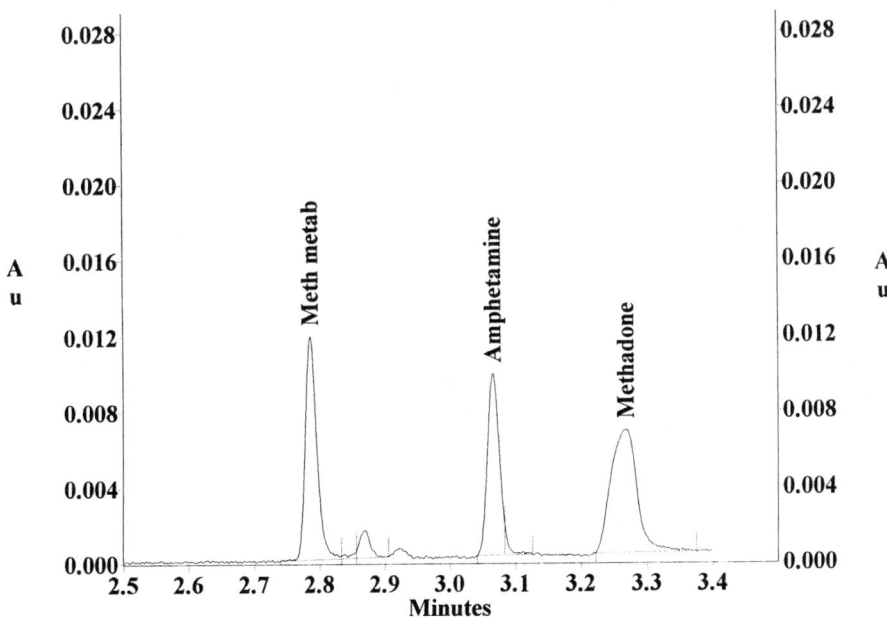

Fig. 5. Electropherogram of a urine from a drug abuser receiving methadone. Sample loading approx one-sixth of **Fig. 4.**

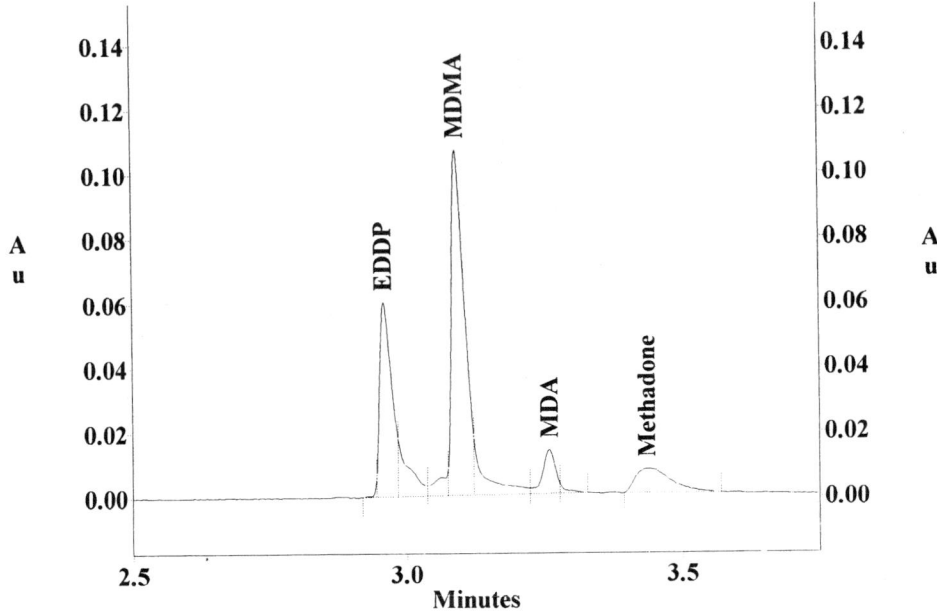

Fig. 6. Urine containing methadone metabolite, MDMA (Ecstasy), MDA, and methadone.

migration times, but fortunately have significant differences in their spectra (*see* **Fig. 7A**).

3. Amphetamine and methamphetamine have identical spectra, but different migration times. MDMA, MDA, and MDE also have identical spectra, but are different from amphetamine (*see* **Fig. 7B**).

4. Notes

1. The pH of the buffer is critical to maintaining good separation of all the amphetamines. The pH of the borax solution is 9.27; this can be used to set the pH meter prior to adjusting the buffer to pH 9.85.
2. Not all basic drugs are sufficiently ionized at pH 4.2 to be retained during the methanol wash. Using 0.1 M HCl, instead of the acetate buffer, permits the detection of the cocaine metabolite, benzoylecgonine (migration time approx 4.2 min). However, the electropherogram becomes significantly dirtier, and this is not recommended for routine detection of opiates.
3. A urine volume of 2 mL gives approximate sensitivities of 0.2 mg/L for opiates, and 0.1 mg/L for amphetamines and methadone. For greater sensitivity, the urine volume can be increased up to 5 mL. This can be useful for dilute urines.
4. It is important not to dry the columns by letting air be drawn through. It is recommended that a system is used that allows vacuum control of individual cartridges.

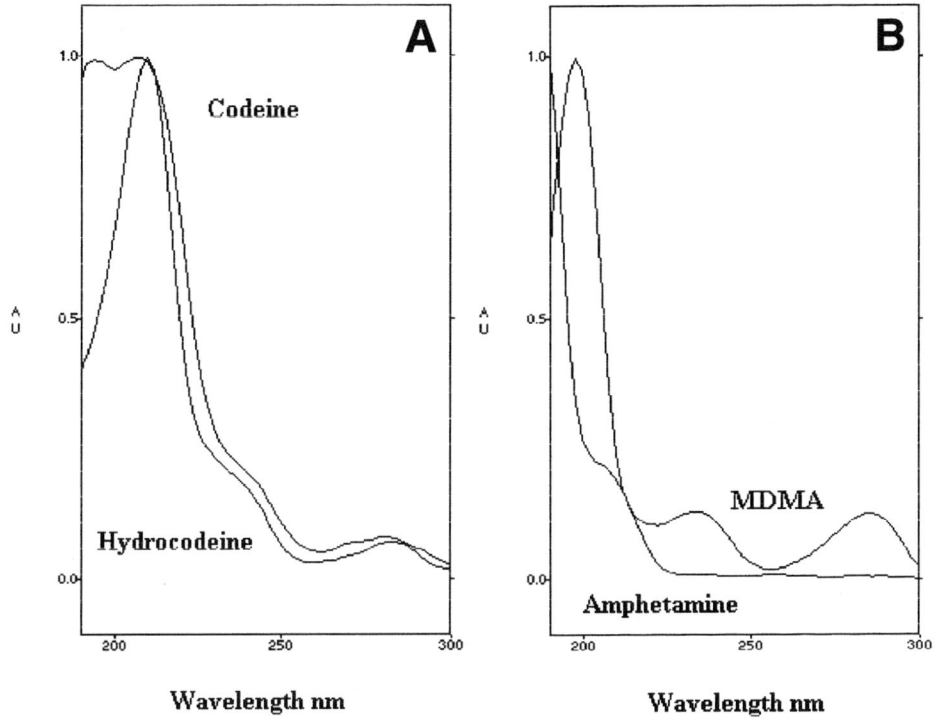

Fig. 7. Diode array scans: **(A)** codeine and hydrocodeine; **(B)** MDMA and amphetamine.

5. The evaporation stage is critical. The opiates are relatively insensitive to temperature, but amphetamines are volatile and great care must be taken with this step. Ideally, a low temperature evaporation (40°C), using nitrogen, should be used. To speed the process, higher temperatures can be employed, up to 70°C, or even 80°C, can be used, provided the sample vessel is removed from the heat as soon as evaporation is complete. Failure to do this can result in complete loss of amphetamine.

6. This is an important step. Repeated washing of the bottom and sides of the tube is essential to ensure that all the residue is redissolved.

7. This applied voltage typically produces a current of 70 µA, and generates much heat. It is very important that the electrophoresis system has an efficient capillary cooling system, otherwise the acetonitrile in the buffer can boil. If problems occur, a lower voltage should be used.

8. Electrokinetic loading gives a considerable increase in sensitivity, compared with pressure loading. Much larger amounts can be loaded without loss of peak resolution.

9. When swapping from one method to another, flush the capillary with 0.2 *M* NaOH for 5 min, followed by water for 1 min. Run the first sample of the next batch twice to equilibrate the capillary.

10. This buffer system will, in addition to opiates, separate methaqualone, phenobarbital, and secobarbital (these all run later than codeine). However, these drugs are lost during the sample preparation. By modifying the preparation, these drugs can be detected, but not at the same time as the opiates.

11. A solvent extraction method, such as Toxi-lab A, yields a sample that contains opiates, benzoylecgonine, methaqualone, barbiturates, methadone, amphetamines, and EDDP. However, this is cumbersome for large numbers of samples, and migration times for amphetamines are erratic.

12. As the methadone concentration rises, the peak produced broadens noticeable. This effect can be seen in **Figs. 5** and **6**. With very high concentrations, the peak can spread off the end of the electropherogram, but it can still be identified by its spectrum.

References

1. Widdop, B. and Caldwell, R. (1991) Operation of a hospital laboratory service for the detection of drugs of abuse, in *Analysis of Drugs of Abuse* (Gough, T. A. ed.), Chester, UK, 429–452.

2. Paul, B. D., Mell, J. R., Mitchell, J. M., Irving, J., and Novak, A. J. (1985) Simultaneous identification and quantitation of codeine and morphine in urine by capillary gas chromatography and mass spectroscopy. *J. Anal. Toxicol.* **9,** 222–226.

3. Cone, E. J. and Darwin, W. D. (1992) Rapid assay of cocaine, opiates and metabolites by GC-MS. *J. Chromatogr. B. Biomed. Appl.* **580,** 43–46.

4. Stead, A. H., Gill, R. Wright, T., Gibbs, J. P., and Moffat, A. C. (1982) Standardised thin-layer chromatographic system for the identification of drugs and poisons. *Analyst* **107,** 1106–1168.

5. Sevenson, J. O. (1986) Determination of morphine, morphine-6-glucuronide and normorphine in plasma and urine with high performance liquid chromatography and electrochemical detection. *J. Chromatogr.* **375,** 174–178.

6. Zoer, J., Virgili, P., and Henry, J. A. (1987) High performance liquid chromatographic assay for morphine with electrochemical detection using an unmodified silica column with a non-aqueous ionic eluent. *J. Chromatogr.* **382,** 189–197.

7. Trenerry, V. C., Robertson, J., and Wells, R. J. (1995) Analysis of illicit amphetamine seizures by capillary electrophoresis. *J. Chromatogr. A.* **708,** 169–176.

8. Lurie, I. S., Chan, K. C., Spratley, T. K., Casale, J. F., and Issaq, H. J. (1995) Separation and detection of acidic and neutral impurities in illicit heroin via capillary electrophoresis. *J. Chromatogr. B. Biomed. Appl.* **669,** 3–13.

9. Walker, J. A., Krueger, S. T., Lurie, I. S., Marche, H. L., Newby, N. (1995) Analysis of heroin drug seizures by micellar electrokinetic capillary chromatography (MECC). *J. Forensic Sci.* **40,** 6–9.

10. Wernly, P., Thormann, W., Bourquin, D., and Brenneisen, D. (1996) Determination of morphine-3-glucuronide in human urine by capillary electrophoresis

amd micellar electrokinetic capillary chromatography. *J. Chromatogr.* **616,** 305–310

11. Tagliaro, F. Poiesi, C., Aiello, R., Dorizzi, R., Ghielmi, S., and Marigo, M. (1993) Capillary electrophoresis for the investigation of illicit drugs in hair: determination of cocaine and morphine. *J. Chromatogr.* **638,** 303–309

12. Molteni, S., Caslavska, J., Alleman, D., and Thormann, W. (1994) Determination of methadone and its primary metabolite in human urine by capillary electrophoretic techniques. *J. Chromatogr. B. Biomed. Appl.* **658(2),** 355–367

19

Steroid Analysis by Micellar Electrokinetic Capillary Chromatography

Amin A. Mohammad, John R. Petersen, and Michael G. Bissell

1. Introduction

From the clinical perspective, steroids have always held a great deal of interest, since they are highly specific in their function. They do not have a general or systemic effect, but instead regulate specific physiological functions, such as sex differentiation, fetus implantation and growth, electrolyte balance, menstrual cycles, and muscle and bone development. Many disorders have been identified as being caused by under- or oversecretion of steroids, i.e., Addison's disease, Cushing's syndrome, hirsutism and virilism, adenomas, congenital adrenal hyperplasia, acromegaly, Liddle's syndrome, hypertension, and so on *(1)*. Diseases resulting from steroid imbalance usually result from the cumulative effect of one or more steroids. To get a better understanding of the pathophysiology resulting from steroid imbalances, the measurement of a profile of steroids is potentially more beneficial than measuring a single steroid. Fiet et al. *(2)*, who used a profile of eight steroids to gain a better understanding of hirsutism and acne in women, demonstrated a good example of this type of study.

Determination of steroids in serum has always been an analytical challenge, because of the low concentrations and structural similarity of the many different steroids. Many methodologies, such as gas chromatography (GC), gas chromatography–mass spectrometry (GC–MS), and high-performance liquid chromatography (HPLC), have been used to separate and quantitate these compounds *(6–8)*. All these methods are sensitive and specific, however, all require extraction, preconcentration, and, sometimes, derivatization, which extend the turn-around-time (TAT), thus limiting their use in the routine examination of patients.

From: *Methods in Molecular Medicine, Vol 27: Clinical Applications of Capillary Electrophoresis*
Edited by: S. M. Palfrey © Humana Press Inc., Totowa, NJ

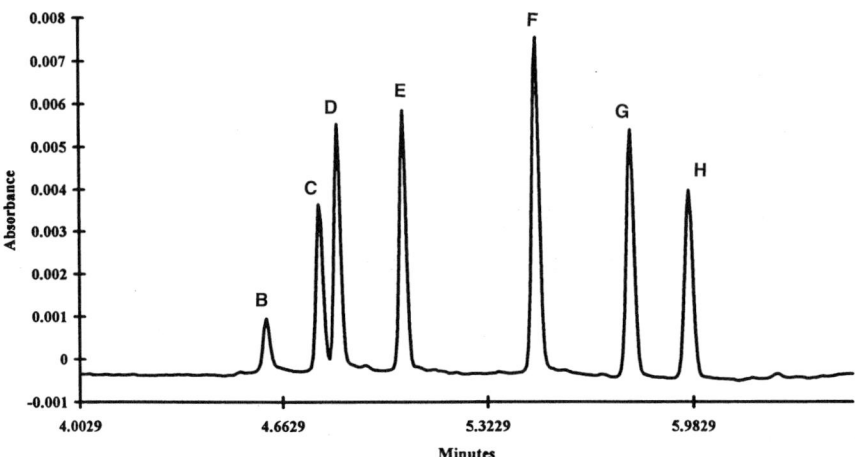

Fig. 1. Spiked serum sample. The steroid peaks are labeled as follows. (*See also* **Fig. 3**.)

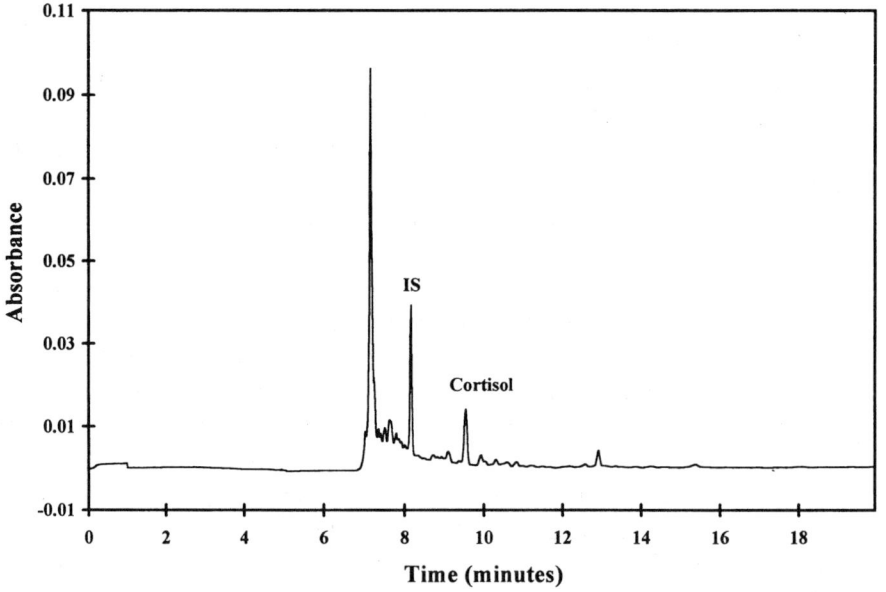

Fig. 2. Urinary free cortisol.

At present, radioimmunoassay (RIA) is the preferred choice for measurement of steroids. Although extremely sensitive, antibody specificity is a problem. For many steroid determinations, enhanced specificity requires extraction and partial purification by chromatographic techniques to remove crossreacting substances. Generally, only a single steroid can be measured in an RIA

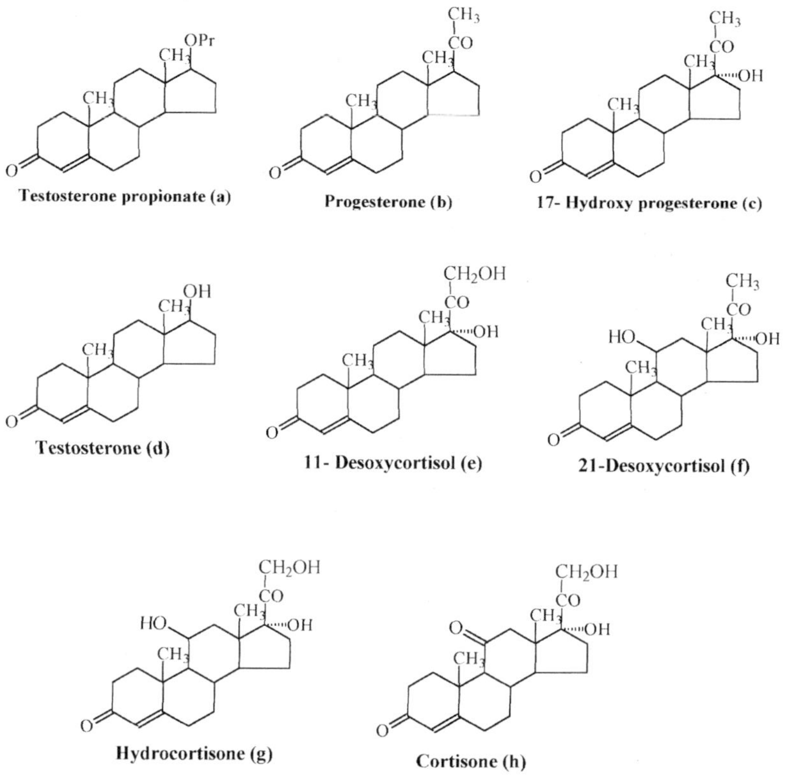

Fig. 3. Steroid structure.

assay. To obtain a profile, multiple assays are required, increasing the TAT and limiting the availability of this application to a few specialized laboratories. The existing methodologies for steroid determination are also costly, limiting their use in the routine examination of patients.

Ideally, a methodology suitable for the quantitation of steroids in clinical samples should meet as many of the following criteria as possible: high sensitivity (detection limits 0.01–10 ng/mL range), high specificity (ability to detect and quantitate steroids in the presence of other structurally similar steroids), minimal derivatization and/or prior sample preparation, fast TAT for high sample throughput, relative low cost potential for automation, and small sample volume (<1 mL of plasma or serum). No method currently available today meets all of these requirements.

Capillary electrophoresis (CE), with its high efficiency, resolution, rapid analysis, and minimal sample and solvent requirements, is one of the more exciting analytical methods to be developed in recent years. CE is an instru-

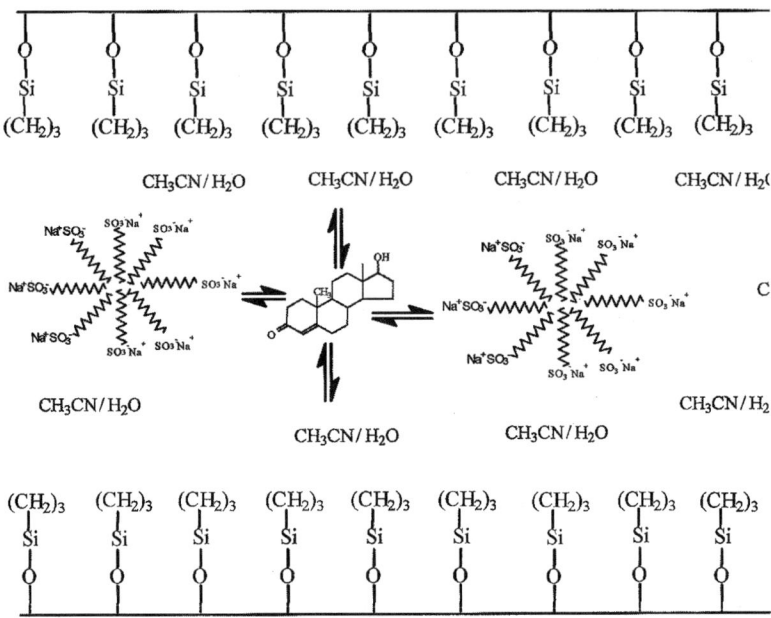

Fig. 4. Mechanism of steroids separation on a neutral capillary.

mental approach to separation by electrophoresis, whose greatest advantage is its diverse application range. It has proved useful for separations of compounds such as chiral drugs, amino acids, vitamins, inorganic ions, organic acids, pesticides, dyes, surfactants, peptides and proteins, carbohydrates, olignucleotides, and DNA restriction fragments. On the simplest level, CE is an application of the principles of chromatography that employs electric current to pull ions of different size and charge through a capillary. The ions are then separated by the speed at which they emerge from the capillary. The speed of the ions depends upon the type and size of the charge, temperature of the capillary column, pH and ionic strength of the buffer, and the viscosity of the buffer. CE has the potential of meeting the criteria for clinical steroid assays because of its high separation efficiency (resolution ~200,000 plate number), low sample and reagent volume, and fast analysis time.

The following discussion describes two typical CE procedures that can be applied to the determination of steroids in urine and serum.

2. Materials

2.1. Instrumentation

1. CE analysis is performed on a Beckman P/ACE 5010 (Beckman, Fullerton, CA) equipped with System Gold software for data analysis.

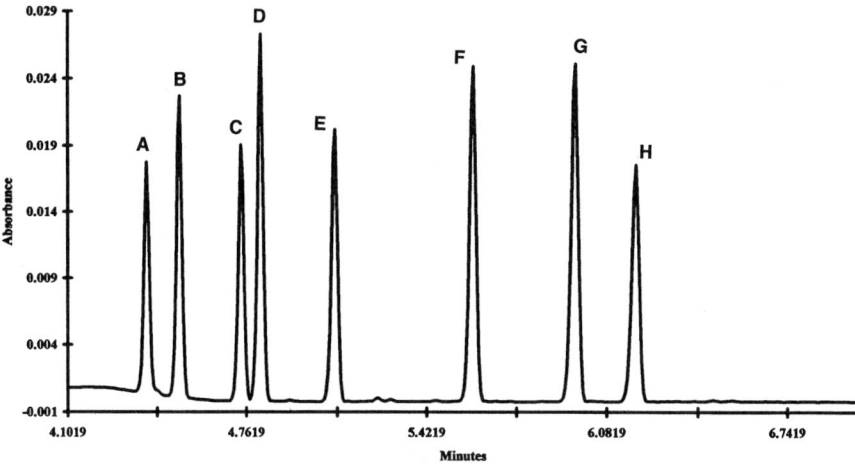

Fig. 5. Separation of eight steroids on a neutral capillary.

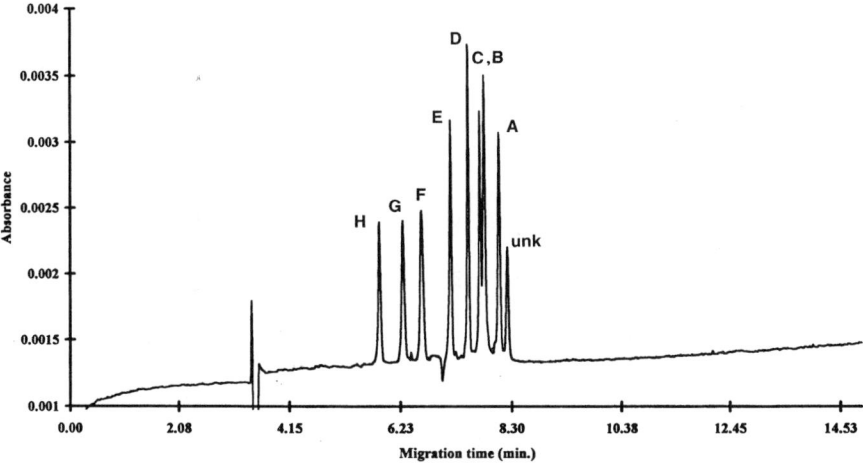

Fig. 6. Separation of steroids on a fused silica capillary. (Run buffer: 5 m*M* SB3-12, 10 m*M* SDS, 25 m*M* bicine, pH 9.0.)

2. Preconcentration of urine samples is achieved by using 3 *M* Empore extraction disk cartridges (Fisher). The disks are C-18 modified 12-mm fused-silica particles, immobilized on an inert matrix of polytetrafluoroethylene (PTFE) fibrils (7 mm diameter, 0.5 mm thick), secured in 3-mL polypropylene columns.
3. Serum ultrafiltrate is collected by filtration through an Amicon micropartition system (MPS-1, Amicon, Beverly, MA).
4. Chemicals: 4-Pregnene-11b, 17,21-triol-3, 20 dione 21-acetate (cortisol acetate), 21-sulfate (cortisol sulfate), 21-glucosiduronate (cortisol glucuronides) are purchased

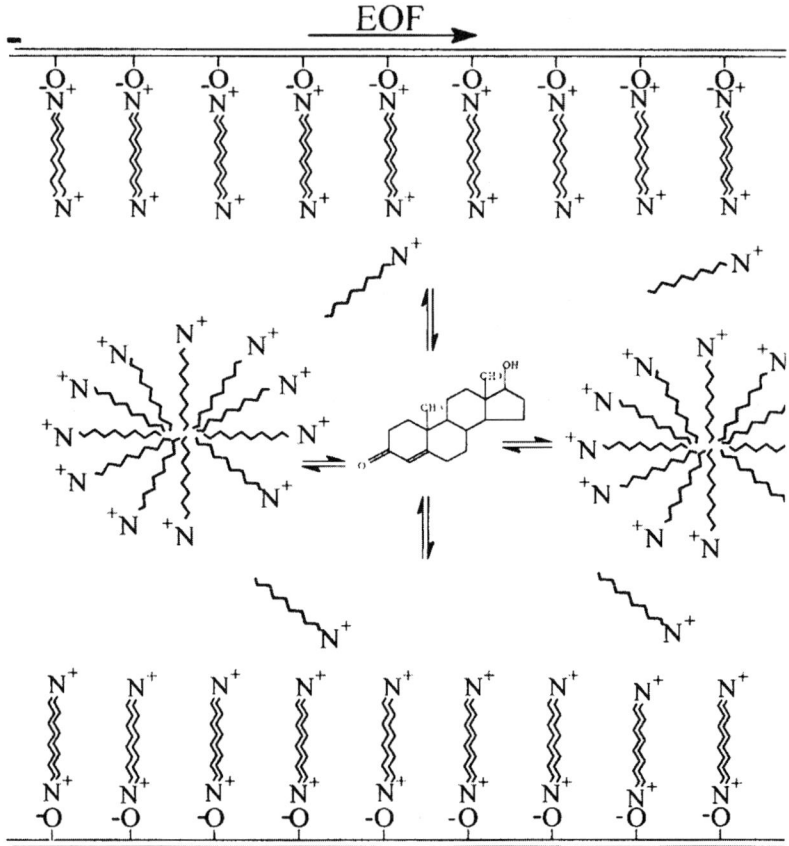

Fig. 7. Mechanism of separation using DTAB as surfactant.

from Steraloids (Wilton, NH). Cortisol (hydrocortisone), corticosterone, testosterone, cortisone, testosterone propionate, 17-OH progesterone, progesterone, 4 pregnene-17, 21 diol 3,20 dione (21-desoxycortisol), 4 -pregnene-17, 11 diol 3,20 dione (11-desoxycortisol) (all 99.9% pure) are purchased from Sigma (St. Louis, MO). Dissolving the respective steroids in 200-proof ethanol makes steroid stock solutions (1 mg/mL).

5. Stock morpholinoethane sulfonic acid (MES) buffer: 100-mM MES, pH 6.0.
6. Electrophoresis buffer: 100 mM/L SDS, 20% acetonitrile (v/v), and 20 mM/L MES buffer, pH 6.0.

3. Methods

3.1. Serum Samples

1. Dilute 0.5 mL spiked serum to 1 mL with 50 mM/L acetate buffer, pH 4.5, and mix by vortexing for 1 min.

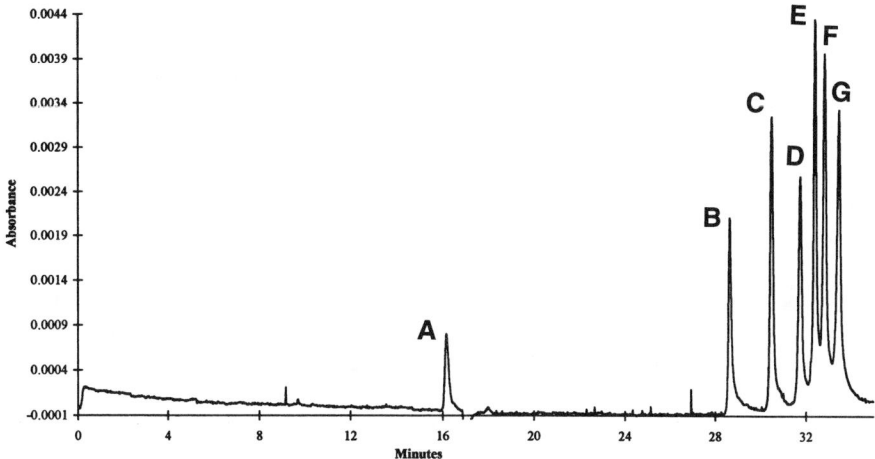

Fig. 8. Separation of steroids using a buffer containing DTAB and TOPO: 50 m*M* DTAB, 5.2 m*M* TOPO, 10 m*M* phosphate, pH 7.4. (Buffer: 50 m*M* DTAB + 5.2 m*M* Topo + 10m*M* phosphate, pH7.4. Analytes: (**A**), mesityl oxide; (**B**), cortisone; (**C**), hydrocortisone; (**D**), 17-deoxycorticosterone; (**E**), testosterone; (**F**), dimethyl testosterone; (**G**), testosterone propionate).

2. Collect an ultrafiltrate (0.4 mL) by filtration through an Amicon micropartition system (MPS-1, Amicon). The resulting ultrafiltrate is directly injected into the CE apparatus for MECC without any further treatment (*see* **Notes 1** and **2**).

3.2. Urine Samples for Determination of Urinary Free Cortisol

1. Preconcentrate urine (10 mL), spiked with 100 mg/dL corticosterone (Internal standard; [IS]) under vacuum using SPE cartridges, preconditioned with 250 μL methanol and washed with 1 mL deionized water (DIW) (*see* **Note 3**).
2. After passing the respective spiked urine samples, wash the disks twice with 1 mL 10% acetone (acetone:water v/v), followed by 1 mL of DIW.
3. Elute the steroids using 80 μL acetonitrile, followed by 320 μL 10 m*M* SDS, and inject them into a neutral capillary for micellar electrokinetic capillary chromatographic separation.

3.3. Capillary Electrophoresis

Micellar electrokinetic capillary chromatography (MECC) is used *(7)* for the separation and detection of urinary free cortisol (*see* **Notes 4–6**).

1. MECC is performed on a 37 cm, 50 μm (id) × 375 mm (od) neutral eCAP capillary tube (Beckman, Brea, CA).
2. Samples are injected under high pressure for 20 s, and the capillary temperature is maintained at 16 ± 0.1°C. Fixed wavelength UV detection at 254 nm, and 10 kV

voltage with reversed polarity is employed for all final separations. Separation is achieved in a neutral eCAP capillary by partitioning steroids between a pseudostationary organic phase (acetonitrile) and an SDS micellar phase.

3. **Figures 1** and **2** *(as shown on p. 178)* show typical electropherograms for determination of steroids in serum and urine respectively.

4. Notes

1. Analysis of steroids in body fluids, especially serum and urine, plays an important role clinically in the differential diagnosis of various inborn errors of steroid metabolism. Congenital adrenal hyperplasia (CAH) is a group of autosomal recessive disorders involving the adrenal glands, in which the primary defect is a deficiency of one or more enzymes involved in the biosynthesis of cortisol. The three principal steroids that are measured during the differential diagnosis of this disease are 17-hydroxy progesterone, 21-deoxycortisol, and 11-deoxycortisol. In CAH the levels of these steroids are elevated in serum, and can reach values as high as 22,000 ng/dL for 17-hydroxy progesterone. Abnormal levels of these steroids are therefore easily detected. The challenge is in detecting normal levels, to differentiate between diseased and nondiseased populations. The maximum normal levels of these steroids in serum are 138 ng/dL and 155 ng/dL for 17-hydroxy progesterone and 11-deoxycortisol, respectively *(15)*. Although MECC can rapidly separate these structurally similar steroids, detection of normal serum levels is not currently possible with the conventional UV absorbance detector available with most CE instruments *(16)*.

   Because of the lack of fluorescence in the steroid nucleus, the cyclopentanoperhydrophenanthrene ring, absorbance is the only mode of detecting these steroids directly in serum with minimal sample preparation. The lower limit of detection for these compounds by MECC is 0.05 mg/dL *(16)*. Thus, a 5000-fold preconcentration is needed to detect normal serum levels. In addition to this limitation in sensitivity, other problems associated with CE include capillary occlusion and related matrix effects. Most body fluids represent a complex matrix with high levels of proteins, and have a tendency to precipitate inside a capillary. These proteins also have high affinity for fused silica, and, over multiple injections, tend to coat the capillary surface, resulting in variable migration times. Therefore, sample preparation plays a very important role in the development of a robust CE assay for the determination of steroids.

2. The most important consideration in serum CE analysis is the removal of serum protein. As mentioned in **Note 1**, samples loaded with proteins can cause capillary clogging. There are a number of ways protein removal can be achieved. One of the simplest is precipitating protein by acidifying the serum. Sulfosalicyclic and trichloroacetic acid (TCA) are the most commonly used precipitants. For this application, TCA is preferred over sulfosalicyclic acid, because of its minimal absorbance near the 254-nm wavelength, which is routinely used for detecting steroids in CE. One of the main problems with this mode of precipitation is sample dilution. This occurs at two stages: once during the addition of acid, and

then again during the process of neutralization. Thus, the low levels of steroids in serum are further lowered by addition of acid and base. An alternative approach is ultrafiltration of serum acidified to around pH 4.0 with acetate buffer through an Amicon micropartition system (MPS-1, Amicon). This is relatively more expensive, but easier to use, and typically involves addition of approx 1 mL serum into the filter assembly, and centrifugation of the assembly for 30 min at 3000g. The ultrafiltrate collected can then be directly injected into the capillary.

3. Sample preparation for urine is much easier, because of the inherently cleaner matrix it represents. In addition, with urine, the relatively large sample size also allows the use of preconcentration techniques, such as solvent and/or solid-phase extractions. Solid-phase extractions can be done by two methods: using C-18 or C-8 extraction cartridges, or, alternatively, by using a solid-phase microextraction disk manufactured by 3 *M*. In the authors' experience the 3 *M* Empore microextraction disk is superior to the cartridge, with recoveries typically ranging from 89–95% (V. R. Lokinendi et al., unpublished data). In addition to excellent extraction efficiencies, they require a relatively low volume of solution to elute the adsorbed steroid, allowing concentration of the steroids in the sample. Disk conditioning is of utmost importance. Usually two washes with methanol and water are sufficient. Use of unconditioned disks results in poor recovery. Application of mild vacuum is recommended during washing and elution stages, because it speeds up the process, and prevents the elution of impurities.

4. MECC is a unique mode of CE, because it can separate neutral, as well as charged, molecules. In MECC, ionic or neutral surfactants are added to the running buffer to form micelles, which are dynamic aggregates that are roughly spherical in shape. They are formed when the concentration of the surfactant reaches the critical micelle concentration (CMC), creating equilibrium between the monomeric surfactant and micelle in solution. Micelles have a three-dimensional structure, with the hydrophobic moieties of the surfactant in the interior and the charged moieties at the exterior. The unique characteristic of micelles is that they provide a microenvironment that is distinctly different from the bulk solvent. This hydrophobic microenvironment provides sites of interaction for solutes in or on the micellar aggregates, thus enhancing the solubility of insoluble nonpolar compounds in aqueous media. The more polar compounds remain in the bulk aqueous medium; nonpolar solutes are partitioned into the micelle. The stronger the interaction, the longer the solutes migrate with the micelle, thereby separating the various components of a mixture based on their partition coefficient. The selectivity of MECC can be controlled by the choice of surfactant, and also by the addition of modifiers to the buffer. As seen in **Fig. 3**, *(on p. 179)* most of the steroids are neutral, highly lipophilic moieties, whose separation inherently requires partitioning between two mediums, and therefore MECC is the only CE mode that can be used to separate them.

5. In MECC, using anionic, cationic, or zwitterionic surfactants can facilitate the separation of different steroids. The key consideration is that the combination of capillary and buffer used for electrophoretic separation should promote parti-

tioning of steroids. By themselves, cationic or anionic surfactants, with either fused-silica or coated capillaries, are unable to resolve the different steroids in the mixture. In the authors' experience, addition of an organic modifier or use of mixed micelles is a necessity for separation. With anionic surfactants such as sodium dodecyl sulfate (SDS), resolution can be achieved by either adding organic modifiers, such as bile salts *(9–10)*, or by using a coated capillary with acetonitrile added to the run buffer, to promote partitioning of steroids between SDS micelles and the hydroorganic phase *(11)*. **Figure 4** *(as shown on p. 180)* shows the basic mechanism behind the separation of steroids, using SDS and an organic modifier such as acetonitrile. Depending on the type of capillary used, the migration order of steroids can be predicted and controlled by using an appropriate buffer combination. In neutral capillaries, such as the Beckman eCAP, the surface charges are neutralized, so migration of steroids is not caused by electroosmotic flow (EOF), but by the electrophoretic mobility of the SDS micelles. Thus, testosterone propionate, which is more lipophilic, partitions readily into the micelles, and is detected first; the more polar cortisone is seen in the end of the run shown in **Fig. 5** *(as shown on p. 181)*. Alternatively, an anionic surfactant can be used in combination with a zwitterionic surfactant, such as N-dodecyl-N, N-dimethyl-3-ammonio-1-propanesulfonate (SB3-12), on fused silica to promote steroid separation *(12)*. The mixed micellar systems offer the advantages of high precision, low Joule heating, and better control over selectivity. This is usually accomplished by varying the zwitterion: SDS ratio. However, decreased separation efficiency (theoretical plates) and increased minimum detection limits offset these advantages. The migration order of steroids is reversed when a combination of SDS and SB3-12 buffer (pH 9.0) is used in a fused-silica capillary, as depicted in **Fig. 6** *(as shown on p. 181)*. In this system, the EOF is toward the cathode, with the SDS/SB3-12 micelles migrating in the opposite direction. Thus, under these conditions, the more lipophilic testosterone propionate is detected last, and the more polar cortisone is detected first.

6. Buffer composed of cationic surfactant, with trioctylphosphine oxide (TOPO) added as the organic modifier, and a fused silica capillary *(13)*, can also affect separation of clinically important steroids. An interesting feature of cationic surfactant is the concentration-dependent reversal in direction of the EOF. This phenomenon was first described by Tsuda *(14)*, and is caused by the reversal of sign of the zeta potential at the fused-silica wall when a surfactant, such as DTAB, is substituted for SDS, as shown schematically in **Fig. 7** *(as shown on p. 182)*. It involves the formation of a bilayer caused by electrostatic attraction between the positively charged amine head group of the cationic surfactant and the negatively charged silica capillary surface. The bilayer formed with the cationic surfactant is a dynamic bilayer phase that acts similar to a C18 bonded phase in reversed-phase, high-performance liquid chromatography (RP-HPLC) *(13)*. Thus, the polar steroids are seen first, followed by the more lipophilic ones, as shown in **Fig. 8** *(as shown on p. 183)*. The separation of steroids in this system is determined by three partition equilibria: between the micellar and the aqueous phase, between

the micellar and dynamic bilayer phase, and between the aqueous phase and the dynamic bilayer phase.

References

1. Greenspan, F. S. (ed.) (1991) *Basic and Clinical Endocrinology*, 3rd ed. Appleton and Lange, CA.
2. Fiet, J., Gosling, J. P., Soliman, H., Galons, H., and Vexiau, P. (1994) Hirsutism and acne in women: coordinated radioimmunoassays for eight relevant plasma steroids. *Clin. Chem.* **40,** 2296–2305.
3. Bosworth, N. and Towers, P. (1989) Scintillation proximity assay. *Nature* **341,** 167–168.
4. Gueux, B., Fiet, J., Pham-Huu-Trung, Villette, J. M., Gourmelen, M., Galons, H., et al. (1985) Radioimmunoassay for 20-deoxy cortisol: clinical application. *Acta Endocrinal (Copenhagen)* **108,** 537–544.
5. Fiet, J., Villette, J. M., Galons, H., Boudou, P., Burthier, J. M., and Hardy, N. (1994) Application of new highly sensitive radioimmunoassay for plasma 21-deoxycortisol to the detection of steroid 21-hydroxylase deficiency. *Ann. Clin. Biochem.* **31,** 51–64.
6. Marx, J. L. (1978) Estrogens: hormones link to cancer disputed. *Science* **202,** 1270–1271.
7. Rudiger, H. W., Haenigsch, F., Metzler, M., Oesch, F., and Glatt, H. R. (1979) Metabolites of diethylstilboesterol induced sister chromatid exchange in human cultured fibroblasts. *Nature* **281,** 392–394.
8. Donike, M. and Zimmermann, J. (1980) Zur darstellung von Trimethylsilyl-, tri-ethysilyl- und tert-butyldimethylsilyl-enolthern von ketosteroiden fr gas-chromato-graphische und massenspektrometrische untersuchungen. *J. Chromatogr.* **202,** 483.
9. Nishi, H., Fukuyama, T., Matsuo, M., and Terabe, S. (1990) Separation and deter-mination of lipophilic corticosteroids and benzothiazepin analogues by micellar electrokinetic chromatography using bile salts. *J. Chromatogr.* **513,** 279–295.
10. Bumgarner, J. G. and Khaledi, M. G. (1996) *J. Chromatogr. B.* **674,** 275.
11. Mohammad, A. A., Bissell, M. G., and Petersen, J. R. (1995) Rapid profiling of clinically relevant steroids by micellar electrokinetic capillary chromatography. *J. Capillary Electrophoresis* **2,** 105–110.
12. Valbuena, G. A., Rao, L. V., Petersen, J. R., Okorodudu, A. O., Bissell, M. G., and Mohammad, A. A. (1997) Anionic-zwitterionic mixed micelles in micellar electro-kinetic separation of clinically relevant steroids. *J. Chromatogr. A.* **781,** 467–474.
13. Mohammad, A. A., Petersen, J. R., and Bissell, M. G. (1995) Micellar electroki-netic capillary chromatographic separation of steroids in urine by trioctylphos-phine oxide and cationic surfactant. *J. Chromatogr. B.* **674,** 31–38.
14. Tsuda, T. (1987) *J. High Resolut. Chromatogr. Commun.* **10,** 622.
15. Tietz, N. W. (ed.) (1995) *Clinical Guide to Laboratory Tests*, 3rd ed. Saunders, Philadelphia, PA.
16. Mohammad, A. A., Petersen, J. R., and Bissell, M. G. (1995) Micellar electroki-netic capillary chromatography to separate steroids that are increased in congeni-tal adrenal hyperplasia. *Clin. Chem.* **41,** 1369–1370.

20

Determination of Polyamines by Capillary Electrophoresis

Yin Fa Ma, Qingnan Yu, and Bingcheng Lin

1. Introduction

The polyamines, mainly putrescine (PU), spermidine (SPD), and spermine (SPM) (**Fig. 1**), are low molecular mass aliphatic amines that exist in all living organs, and play important roles in cell growth and differentiation *(1)*. The intracellular concentrations in many cell types are in the submillimolar range, and the maximum increase is less than twofold over the course of a cell cycle *(2)*. It is well known that ornithine-derived polyamines are implicated in a variety of cell functions involving DNA replication, gene expression, and protein synthesis *(3,4)*. Recent studies demonstrated that ornithine-derived polyamines play roles in potassium depolarized stimulation of synaptic function *(5)*, determination of tumor malignancy *(6)*, intercellular messengers *(7)*, stabilization of membrane structures *(4)*, and regulated exocytosis *(8,9)*.

The regulation of polyamine production is essential, because the overproduction of polyamines is toxic to cells, and facilitates cell death by oxidative mechanisms *(10,11)*. Polyamine researches demonstrated that tumor cells contain a much higher concentration of polyamines, and patients with many types of cancers have an elevated urine polyamine concentration *(12)*. After that, many reports claimed that polyamine concentrations in different kinds of tumors vary widely *(13–16)*. So far, though specific malignancy changes in polyamine profile for different cancers have not been observed, it has been proved that polyamine concentrations respond to effective medical therapy, and that they might be used as indicators for posttherapeutic relapse of tumors *(17)*. Therefore, a simple, rapid, and sensitive method of polyamine determination could be very useful in studying the biological role of polyamines, and in predicting the effectiveness of medical therapy on cancers.

From: *Methods in Molecular Medicine, Vol 27: Clinical Applications of Capillary Electrophoresis*
Edited by: S. M. Palfrey © Humana Press Inc., Totowa, NJ

$$NH_2 (CH_2)_4 NH_2$$
Putrescine (PU)

$$NH_2(CH_2)_3NH(CH_2)_4NH_2$$
Spermidine (SPD)

$$NH_2(CH_2)_3NH(CH_2)_4NH(CH_2)_3NH_2$$
Spermine (SPM)

Fig. 1. The three common polyamines.

Traditional methods for polyamine detection are mostly confined to high-performance liquid chromatography (HPLC) *(18–20)*, thin-layer chromatography *(21)*, and, sometimes, gas chromatography *(13)*. All these chromatographic methods, however, require that polyamines be derivatized or labeled before detection, since they have no chromaphore, and cannot be detected with an UV detector. This makes the methods tedious and time-consuming. Capillary electrophoresis (CE) has been one of the fast-growing analytical techniques for the past several years, and it has been rapidly accepted by more and more biomedical and clinical researchers, because of its strong power in separating charged biomolecules with high resolution and extremely small sample requirement *(22,23)*. The basic principles and the applications of CE has been published in several books *(24–27)*. Capillary zone electrophoresis (CZE), with indirect UV detection, has been proved to be a rapid, simple, and sensitive method for determination of polyamines in biosamples such as serum *(28)* and tumor cells *(29,30)*. The indirect detection technique has become a powerful tool in many cases *(31,32)*. The basic principle is that the detector responds to some physical property of the chromaphore in the running buffer. There is a constant background signal generated at the detector when no analytes are present. When the analyte migrates to the detector window, it replaces an equal amount of the chromaphore. Even though the detector does not respond to the analyte, the lower chromaphore concentration at the detector causes a decrease in signal. The analyte can be monitored as a negative signal. The size of the negative peak area is proportional to the analyte concentration at certain concentration ranges.

Polyamines exist in all living organs in two types: unbound, free polyamines; and bound polyamines, which bind to macromolecules, such as DNA, RNA, protein, and phospholipids. To serve different purposes, one might prefer to know either free polyamines, bound polyamines, or total polyamines, which are bound polyamines plus unbound polyamines. Different needs can be easily met, in many cases, by pretreating the biosamples differently before they are loaded onto the CE system for separation and quantitation.

2. Materials

As mentioned in the earlier section, polyamines have no chromophore, hence no UV absorbance, which is required for direct UV detection. Therefore, indirect UV detection is used to monitor polyamines. Here, quinine sulfate is a very good candidate for this purpose. Quinine sulfate has a maximum absorbance at 236 nm, with an absorption coefficient of 34,900 /mol cm. When it is contained in the background electrolyte solution, it gives a stable absorbance at 236 nm wavelength, and being served as the baseline absorbance. When the nonabsorbing polyamines flow through the detection window, reverse peaks are obtained, which makes the detection and quantitation of them possible.

2.1. Background Electrolyte (BGE)

1. Dissolve 329 mg quinine sulfate in a mixture of 20 mL 95% ethanol and 70 mL deionized water.
2. Adjust the pH to 3.0 with 0.1 *M* HCl; the volume is then brought to 100 mL (*see* **Notes 1** and **2**).
3. Degas the solution before use.

2.2. Electrophoresis

1. Fused-silica capillary, 25 μm or 50 μm id and 35–45 cm effective length.
2. The authors use a HPCZE system (Model 3580) from Isco (Lincoln, NE) and collect data with a Datajet Computing Integrator (Spectra-Physics, Mountain View, CA).

2.3. Solutions

1. 50 μ*M* 1,7-diaminoheptane (internal standard) in ice cold deionized water.
2. 50 μ*M* 1,7-diaminoheptane in ice cold 10% trichloroacetic acid (TCA).

3. Methods

The purpose of sample pretreatment is to prepare a polyamine-containing fluid, either free or total polyamine, so that it can be loaded onto the capillary column and successfully run through.

3.1. Serum Pretreatment

1. Add 100 μL serum to 1.0 mL acetone, and remove the denatured proteins by centrifugation at 600*g* for 10 min (*see* **Note 3**).
2. Obtain deproteinized serum by volatilizing the acetone from the superatant.
3. A volume of 72–75 μL deproteinized serum is obtained from the original serum sample. The concentrations of polyamines obtained from the calibration curve need to be converted to 100 μL.

3.2. Preparation of Cell Extracts

3.2.1. Cytosolic Polyamines Preparation

1. Harvest about 5×10^6 cells by centrifugation at 600g for 15 min.
2. Wash the pellet with 0.3 M sucrose.
3. Add 1.0 mL ice-cold deionized water containing 50 μM 1,7-diaminoheptane to the pellet, and hold the sample at 0–4°C for about 2 h.
4. Remove insoluble membranous materials by centrifugation at 60,000g at 0°C for 30 min. Collect the supernatant for injection into the CE column for cytosolic polyamine analysis. This method gives the free polyamine concentration.

3.2.2. Total Intracellular Polyamine Preparation

1. Harvest about 5×10^6 cells by centrifugation at 500g for 15 min.
2. Wash the pellet with 0.3 M sucrose.
3. Add 500 μL 10% ice cold TCA/1,7-diaminoheptane to the pellet.
4. Hold the sample at 4°C for 2 h.
5. Remove precipitated proteins and phospholipid and other insoluble material by centrifugation at 1900g for 15 min.
6. Extract the TCA soluble fraction 6× with 3 vol of ether, to partially remove the TCA and increase the pH to 2.8 (*see* **Note 4**).
7. Inject the solution onto the capillary column for total polyamine analysis.

3.3. Electrophoresis

The loading and running conditions might be slightly different using CE products from different companies. However, the electrophoretic conditions mentioned below for polyamines could be applied to different instruments.

3.3.1. Capillary Pretreatment

1. Fill new capillaries with 0.1 M NaOH solution for 30 min, to clean the column. Then wash the columns with deionized water and background electrolyte. The capillaries are now ready for use.
2. It has been found *(28)* that coated capillary columns work better for analyzing polyamines in serum samples, because they avoid the adsorption of proteins onto the wall of capillary. Better reproducibility is also obtained (*see* **Note 5**).
3. To use a coated capillary, clean a capillary with the same id and effective separation length with 0.1 M HCl and 0.1 M NaOH for 30 min each, in sequence, and then coat with polyacrylamide.
4. Wash the coated capillary with deionized water and background electrolyte before using (*see* **Note 6**).

3.3.2. Separation of Polyamine Standards

1. Use a 60 cm × 25 μm capillary.
2. Sample injection is electrokinetic, using 30 kV for 5 s.

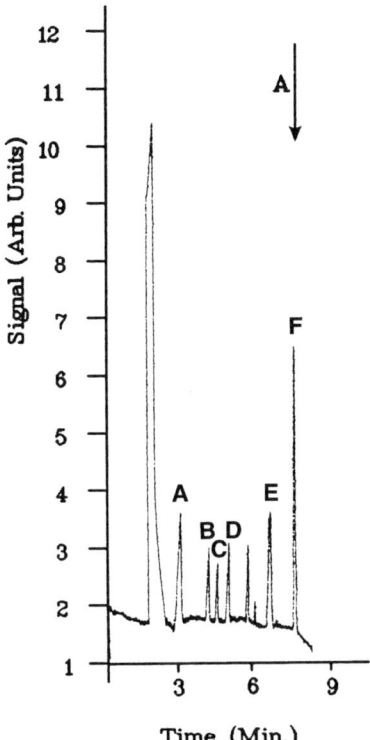

Fig. 2. Separation of three polyamine standards and other co-existing cations in PC 12 cells by HPCZE with indirect photometric detection. A 5-s, 30 kv injection of 50 μm each was followed by electrophoresis at 30 kv on a 60 cm × 25 mm I. S. pretreated column (35 cm to detector). Injection volume was 3 ml. Detection wavelength was 236 nm. Peaks: (**A**) K^+; (**B**) Na^+; (**C**) putrescine; (**D**) 1,7-diaminoheptane; (**E**) spermidine; (**F**) spermine.

3. Separation voltage 30 kV.
4. Detection wavelength 236 nm.
5. Three polyamines and coexisting K^+ and Na^+ in biological samples can be completely separated in less than 10 min. **Figure 2** shows the separation of three polyamines and other commonly existing cations (Na^+ and K^+) added to cell-culture media, together with the internal standard 1,7-diaminoheptane.
6. 1,7-Diaminoheptane, which has not been found in cells, and has physicochemical properties similar to those of polyamines, is used as an internal standard.
7. The linearity of the polyamine detection is over two orders of magnitude for each polyamine, which makes it easier to quantify the polyamines in a wide range of concentrations in different biological samples.

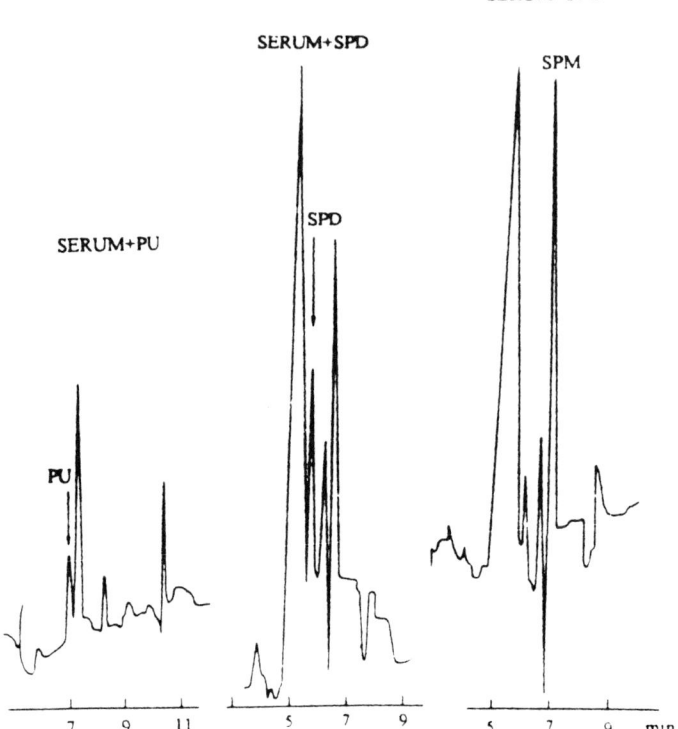

Fig. 3. Electropherograms of serum sample from a normal adult with the addition of PU, SPM and SPD standard separated by CE with indirect UV detection. A 3-s, 5 kv injection was followed by electrophoresis at 9 kv on a 50 μm I. D. polyacrylamide coated column, detected at 236 nm, 0.005 AU.

3.3.3. Determination of Polyamines in Serum Samples

1. **Figure 3** shows the electropherogram of a serum sample from a normal adult, with the addition of PU, SPM, and SPD standards aimed at showing the relative positions of the three common polyamines in serum (*see* **Note 7**).
2. Use a 60 cm × 50 μm polyacrylamide coated capillary.
3. Sample loading is elecktrokinetic, using 5 kV for 3 s.
4. Separation voltage 9 kV.
5. Detection wavelength 236 nm.

3.3.4. Detection of Polyamines in Cell Samples

1. **Figure 4** is an elecropherogram for total polyamine determination in PC12 tumor cell extract.
2. Use a 60 cm × 25 μm capillary.

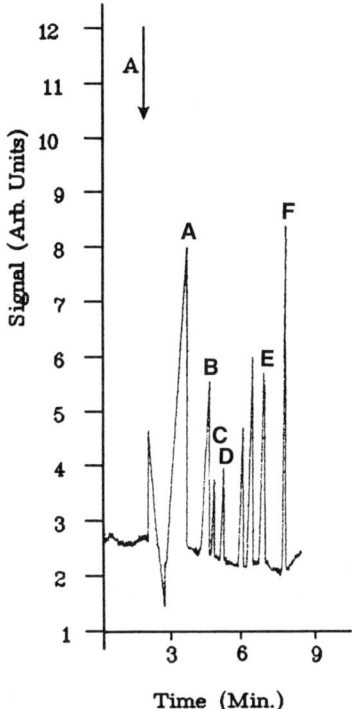

Fig. 4. Separation of threee polyamines and other commonly co-existing cations extracted from 5×10^6 PC-12 cells by HPCZE with indirect photometric detection. Sample was injected at 30 kv for 10 seconds and followed by electrophoresis at 30 kv. Injection volume was 6 nl. Electrophoresis conditions and peak identification are the same as those of **Fig. 2**.

3. Sample injection is electrokinetic, using 30 kV for 5 s.
4. Separation voltage 30 kV.
5. Detection wavelength 236 nm.

3.4. Calibration by External Standards

1. Body polyamine concentrations are mostly in the submillimolar range. A linear response of over two orders of magnitude ($1.0 \times 10^{-3} - 3.0 \times 10^{-6}$ M) for each polyamine makes the quantitation by a calibration curve more convenient.
2. Inject five or six different concentrations within the range of $10^{-3} - 10^{-6}$ M for each polyamine.
3. Construct a calibration curve by plotting peak area as a function of polyamine concentration (*see* **Notes 8** and **9**). Regression equations for each of the curves should be obtained. Use these regression equations for calculating the polyamine concentrations for the biological samples being analyzed.

4. Note that the samples should be analyzed with exactly the same loading and running conditions as the polyamine standards, for highest accuracy.
5. When using a calibration curve method for quantitation, one really has to take the factor of relative recovery of the polyamines during the sample preparation into consideration. If recovery is a major problem, this quantitation technique is not recommended. However, for the tumor cell sample and serum sample preparation mentioned above, the recovery is excellent, and this technique can be used for quick polyamine analysis.

3.5. Calibration with an Internal Standard

1. The use of an internal standard has better precision for quantitative analysis in chromatography and CE, because the uncertainties introduced by sample preparation and sample injection can be corrected.
2. For the quantitation of polyamines, 1,7-diaminoheptane serves as a good internal standard, since it has not been found in either cells nor serum, and it has physicochemical properties similar to those of polyamines.
3. The basic principle is that 1,7-diaminoheptane at a known concentration as a control is added to each polyamine standard, and to all the samples, and goes through all the sample preparation procedures.
4. The ratio of polyamine to internal standard peak area serves as the analytical parameter. A calibration curve is prepared by plotting the ratio of polyamine peak area to internal standard peak area vs the standard polyamine concentration.
5. The polyamine concentration in the sample is found by using the analytical parameter for each polyamine peak and internal standard. Since this method controls the variations in sample preparation and injection, the results will be much more reliable.

4. Notes

1. The preparation of the background electrolyte (BGE) solution is very important for maintaining a good baseline during the separation. Therefore, quinine sulfate should be dissolved in ethanol first. If the quinine sulfate does not dissolve completely, some deionized water can be added. If it still does not dissolve completely, 0.1 M HCl can be added to dissolve the rest of the quinine sulfate and adjust the pH to 3.0 (quinine sulfate dissolves better in acidic solutions). The pH of the BGE is also very critical for the separation of polyamines.
2. During the separation, both BGE vials should be covered with parafilm or other covers, to prevent the evaporation of ethanol. If evaporation happens, the quinine sulfate could be precipitated out. It is suggested that the BGE should be checked frequently, to make sure that it is still in good condition.
3. During the process of serum sample treatment, proteins should be precipitated as thoroughly as possible, so that the adsorption of proteins onto the capillary walls could be minimized.
4. During the cell extraction for the total polyamine determination, the extra TCA should be removed by ether extraction as much as possible, so that a good baseline can be maintained after sample injection.

5. The polyacrylamide coated column will offer a better separation. A nonpoly-acrylamide-coated column can be used, but a longer column needs to be adopted.
6. It is suggested that a capillary column is rinsed with BGE before each run. If it is difficult to obtain the even baseline that one used to, it is time to change to a new capillary.
7. Identification of the polyamine peaks for both serum and cell extract samples is by the standard addition method.
8. When constructing a calibration curve, five repeated runs are needed for every concentration, and average values are used for making the calibration curve.
9. For each installed new column, a new calibration curve should be made for controlling the variations from the column preparation.

References

1. Marton, U. and Morris, D. R. (1987) Molecular and cellular functions of the polyamines, in *Inhibition of Polyamine Metabolism* (McCann, P. P., Pegg, A. E., and Sjoerdsma, A., eds.), Academic, Orlando, FL, pp. 79–106.
2. Sunkara, P. S., Ramakrishna, S., Nishioka, K. and Rao, P. N. (1981) Relationship between levels and rates of polyamines during mammalian cell cycle. *Life Sci.* **28,** 1497–1506.
3. Margan, D. M. L. (1990) Polyamines and cellular regluation: perspectives. *Biochem. Soc. Trans.* **18,** 1080–1084.
4. Marton, U., Pegg, A. E., and Morris, D. R. (1991) Directions for polyarnine research. *J. Cell. Biochem.* **45,** 7–8.
5. Iqbat, Z. and Koenig, H. (1985) Polyamines appear to be second messengers in mediating Ca^{2+} fluxes and neurotransmitter release in potassium-depolarized synaptosomes. *Biochem. Biophys. Res. Commun.* **133,** 563–573.
6. Ernestus, R-I., Rohn, G., Schroder, R., Klug, N., Hossmann, K.-A., and Paschen, W. (1992) Activity of ornithine decarboxylase (ODC) and polyamine levels as biochemical markers of malignancy in human brain tumors. *Act. Histochem.* **42,** 159–164.
7. Feige, J. J., Cochet, C., and Chambaz, E. M. (1992) Potential role of polyamines as intracellular messengers in hormonal regulation of cellular activity, in *Recent Progress in Polyamine Research* (Selmeci, L., Brosnan, M. E., and Seiler, N., eds.), H. Stilman, Boca Raton, FL, pp. 181–189.
8. Hougaard, D. M. and Larsson, L. I. (1986) Localization and possible function of polyamines in protein and peptide secreting. cells. *Med. Biol.* **64,** 89–94.
9. Che, P. (1995) *Effect of Polyamine Depletion on Norepinephrine Metabolism* (M. S. Thesis). Northeast Missouri State University, pp,42.
10. Morris, D. R. (1991) New perspective on ornithine decarboxylase regulation: prevention of polyamine toxicity is the overriding theme. *J. Cell. Biochem.* **46,** 102–105.
11. Packham, G. and Cleveland, J. L. (1994) Ornithine decarboxylase is a mediator of c-myc-induced apoptosis. *Mol. Cell. Biol.* **14,** 5741–5747.
12. Russel, D. H. (1971) Increased polyamine concentrations in the urine of human cancer patients. *Nat. New. Diol.* **29, 223,** 144–145.

13. Russel, D. H. and Russel, S. D. (1975) Relative usefulness of measuring polyamines in serum, plasma, and urine as biochemical markers of cancer. *Clin. Chem.* **21,** 860.

14. Denton, M. D., Glazer, H. S, Zellner, D. C., Smith, F. G. (1973) Gas chromatographic measurement of urinary polyamines in cancer patients. *Clin. Chem.* **19,** 904–907.

15. Loser, C., Folsch, U. R., Paprotny, C., and Creutzfeldt, W. (1990) Evaluation of polyamine concentrations in the colon tissue, serum, and urine of 50 patients with colorectal cancer. *Cancer* **65,** 958.

16. Russel, D. H., Levy, C. C., Schimpff, S. C., and Hawk, I. A. (1971) Urinary polyamines in cancer patients. *Cancer Res.* **31,** 1555.

17. Nishika, K. (1993) Basic mechanisms and clinical approaches cancer research. *Cancer Res.* **53,** 2689.

18. Minocha, S. C., Minocha, R., and Robie, C. A. A. (1990) High perfotmance liquid chromatographic method for the determination of dansyl-polyamines. *J. Chromatogr.* **511,** 177–180.

19. Seiler, N. (1983) Liquid chromatographic methods for assaying polyamines using prechromatographie derivatization. *Methods Enzymol.* **94,** 10.

20. Matsumoto, T. and Tsuda, T. (1990) Polyamine determination in clinical laboratories by high-performance liquid chromatography. *Trends Anal. Chem.* **9,** 292–296.

21. Seiler, N. (1983) Thin-layer chromatography and thin-layer electrophoresis of polymers and their derivatives. *Methods Enzymol.* **94,** 1.

22. Heiger, N. N., Cohen, A. S., and Karger, B. L. (1975) Separation of DNA restriction fragment by high performance capilliary electrophoresis with low and zero crosslinked polyacrylamide using continuous and pulsed electric fields. *J. Chromatogr.* **516,** 33.

23. Novotny, M. V., Cobb, K. A., and Lin, J. (1990) Recent advances in CE of proteins, peptides and amino acids. *Electrophoresis* **11,** 735.

24. Camilleri, P. (1993) *Capillary Electrophoresis, Theory and Practice.* CRC, Boca Raton, FL.

25. Landers, J. P. (1994) *Handbook of Capillary Electrophoresis.* CRC, Boca Raton, FL.

26. Cohen, A. S., Terabe, S., and Deyl, Z. (1996) *Capillary Electrophoresis Separation of Drugs.* Elsevier, New York.

27. Lin, B. C. (1996) *Introduction to Capillary Electrophoresis.* Science Press, Beijing.

28. Zhou, G., Yu. Q., Ma, Y., Xue, J., Zhang, Y., and Lin, B. C. (1995) Determination of polyamines in serum by HPCE with indirect UV detection. *J. Chromatogr. A.* **717,** 345.

29. Ma, Y., Zhang, R., and Cooper, C. L. (1992) Indirect determination of polyamines in biological samples separated by high-performance capillary electrophoresis. *J. Chromatogr.* **608,** 93.

30. Zhang, R., Cooper, C. L., and Ma, Y. (V) Determination of total polyamines in tumor cells by HPCZE with indirect photometric determination. *Anal. Chem.* **65,** 704.

31. Kuhr, W. G. and Yeung, E. S. (1988) Optimization of sensitivity and separation of capillary zone electrophoresis with indirect fluorescence detection. *Anal. Chem.* **60,** 2642.

32. Yeung, E. S. (1989) Indirect detection methods for capillary separations. *Acc. Chem. Res.* **22,** 125.

21

Urinary Oxalate and Citrate

Ross P. Holmes and Martha Kennedy

1. Introduction

The amounts of oxalate and citrate excreted in urine, and their urinary concentrations are important risk factors for the development of calcium oxalate kidney stones (1). The most widely used procedures to estimate these analytes are enzyme-based procedures using commercially available reagent kits (2,3). Components in the urine matrix may interfere with these assays, and some sample cleanup is required to remove them for oxalate analysis. Ion chromatography, although well suited to these determinations (4,5), is less widely used presumably because of long assay times, the need for expensive equipment, and the maintenance costs associated with the procedure. Capillary electrophoresis enables the rapid determination of oxalate and citrate in the same run, as well as the simultaneous measurement of chloride and sulfate. Estimation of these anions is useful for the calculation of relative supersaturations of urine with calcium oxalate and calcium phosphate. The method developed here utilizes indirect absorption to detect anions (6), and relies on the change in absorption observed when oxalate and citrate migrate through the detection window, and displace chromate in the electrolyte. It is important that the chromophoric electrolyte chosen has a migration time similar to that of the anions of interest. Pyromellitic acid is another suitable chromophore that can be used as the electrolyte (7).

2. Materials

1. Electrolyte: The electrolyte is made fresh daily by adding 5 mL 0.2 M sodium chromate to 92.5 mL reagent-grade water, followed by the addition 2.5 mL of 20 mM tetradecyl-trimethylammonium bromide (TMAB) (see **Notes 1** and **2**). The sodium chromate and TMAB stock solutions can be stored indefinitely at

From: *Methods in Molecular Medicine, Vol 27: Clinical Applications of Capillary Electrophoresis*
Edited by: S. M. Palfrey © Humana Press Inc., Totowa, NJ

room temperature without any noticeable deterioration. To avoid interference from small gas bubbles during the run, the electrolyte can be degassed by sonication or vacuum filtration. The Waters Quanta 4000 system (Milford, MA) requires 17 mL electrolyte in each electrode reservoir, this volume may vary, depending on the equipment used.

2. Oxalate and citrate standards: Standards of oxalic acid and trisodium citrate are prepared in 13.0 g NaCl/L, 2.2 g Na_2SO_4/L, and 15.5 g H_3PO_4/L and are diluted 1/100 before use (*see* **Note 3**). A standard curve is generated from solutions containing 20–500 mg/L oxalate and citrate before dilution.

3. An uncoated fused silica capillary, 60 cm in length and 75 μm in id is used, with the detection window set 7–8 cm from the end. A new capillary (*see* **Note 4**) is first washed with 0.1 *M* KOH for 10 min, followed by a 5 min water wash and a 5-min wash with a urine sample diluted 1:100 with water. This wash with urine produces a stable migration time for oxalate and citrate. Without it, the migration time will slowly increase with the first 15–20 urine samples run.

3. Methods

1. Acidify an aliquot of a urine sample by the addition of 10 μL concentrated H_3PO_4 per mL. If acid preservation of the urine is required during its collection, then H_3PO_4, rather than HCl, should be used. The authors have found that acid preservation of 24-h urine collections is normally not required to prevent changes in the urine concentrations of citrate and oxalate. This acidified aliquot can be stored at –20°C, awaiting analysis.

2. The amount of acid added is not sufficient to prevent calcium oxalate crystallization, and samples must be heated at 60°C for 30 min before analysis, to ensure the complete dissolution of crystals.

3. To prepare samples for injection, dilute them 100-fold in water (*see* **Note 5**). Overly concentrated urine samples (24 h vol <1000 mL) can be diluted 1:200 for a better analysis, and overly dilute samples (24 h vol >2500 mL) diluted 1:50 (*see* **Note 3**).

4. The settings used on the Quanta 4000 are a constant current of 25 μA with reversed polarity, a hydrostatic load for 100 s (*see* **Note 6**), an initial wash with 0.1 *M* KOH for 1 min, a wash with 0.1 *M* HCl for 1 min, a wash with electrolyte for 2 min, and a run time of 6 min. Set the detector to 254 nm.

5. Peak areas are determined manually using Millennium software (Waters), and are related to the standard curve (*see* **Note 7**). A comparison of standards and a urine sample is shown in **Fig. 1**.

4. Notes

1. It is important to add the electroosmotic modifier last, to avoid its precipitation from solution.

2. Waters now offers myristyl-trimethylammonium hydroxide as an electroosmotic modifier, to eliminate a negative peak associated with the use of a bromide salt. It is expensive, however, and is only required if an accurate estimate of Cl⁻ is desired.

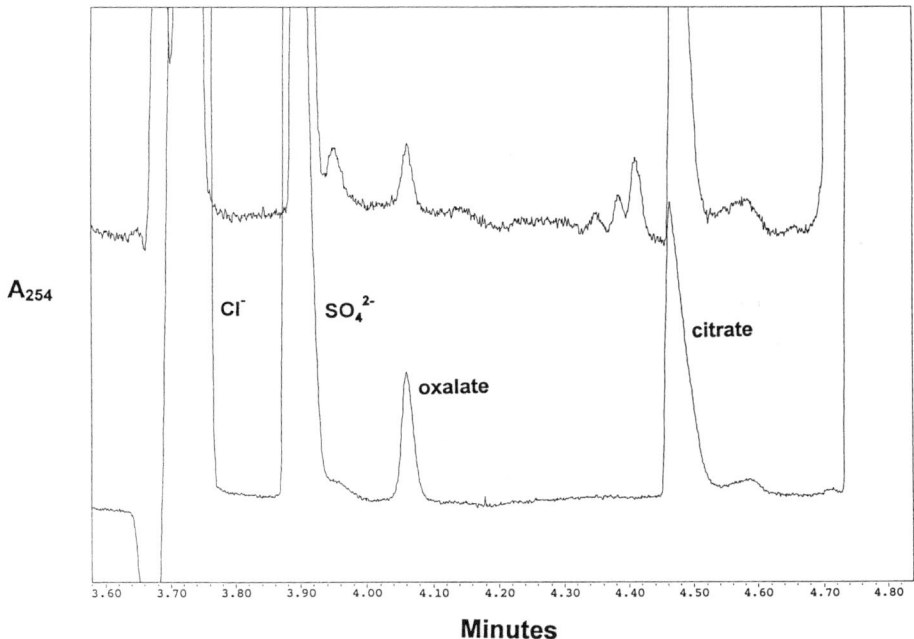

A_{254}

Cl⁻ SO_4^{2-} citrate

oxalate

3.60   3.70   3.80   3.90   4.00   4.10   4.20   4.30   4.40   4.50   4.60   4.70   4.80

Minutes

Fig. 1. Electropherograms of a urine sample (top) and a standard (bottom).

3. It is generally not recognized that the concentration of anions in the sample affects the peak area with indirect absorbance detection and low ionic strength electrophoresis buffers. This necessitates the preparation of standards in an ion mixture mimicking urine. This effect has implications for the analysis of urine, when it is possible for the ionic strength to vary 10-fold.

4. Capillaries may be reused until a substantial drift in the migration times of oxalate and citrate is observed. The authors have usually observed that prolonged washing/stripping of the column does not restore the migration time, and it is advisable to replace the capillary.

5. The dilution of sample with water ensures that there will be a stacking effect, with sharp peaks caused by the high local field strength in the sample zone when the current is applied *(8)*.

6. Although a 100-s load is longer than that generally used in most applications (30–50 s), it gives reproducible results without substantial peak broadening, and the profile obtained is better than that obtained with a 50-s load of a sample diluted 1:50 in water.

7. An internal standard would be very useful. The authors have examined the use of tungstate, molybdate, and fluoride as internal standards, but have not found any of them to be satisfactory.

References

1. Ruml, L. A., Pearle, M. S., and Pak, C. Y. C. (1997) Medical therapy. Calcium oxalate urolithiasis. *Urol. Clin. N. Am.* **24,** 117–133.
2. Warty, V. S., Busch, R. P., and Virji, M. A. (1984) Kit for citrate in foodstuffs adapted for assay of serum and urine. *Clin. Chem.* **30,** 1231–1233.
3. Li, M. G. and Madappally, M. M. (1989) Rapid enzymatic determination of urinary oxalate. *Clin. Chem.* **35,** 1330–2333.
4. Singh, R. P. and Nancollas, G. H. (1985) Determination of urinary citrate by high performance ion chromatography. *Clin. Chem.* **28,** 985–987.
5. Hagen, L., Walker, V. R., and Sutton, R. A. L. (1993) Plasma and urinary oxalate and glycolate in healthy subjects. *Clin. Chem.* **39,** 134–138.
6. Holmes, R. P. (1995) Measurement of urinary oxalate and citrate by capillary electrophoresis and indirect ultraviolet absorbance. *Clin. Chem.* **41,** 1297–1301.
7. Harrold, M. P., Wojtusik, M. J., Riviello, J., and Henson, P. (1993) Parameters influencing separation and detection of anions by capillary electrophoresis. *J. Chromatogr.* **640,** 463–471.
8. Friedberg, M. A., Hinsdale, M., and Shihabi, Z. K. (1997) Effect of pH and ions in the sample on stacking in capillary electrophoresis. *J. Chromatogr.* **781,** 35–42.

22

Plasma Nitrite and Nitrate Determination

Toshiko Ueda, Tsuyoshi Maekawa, and Kazuyuki Nakamura

1. Introduction

The concentrations of nitrite and nitrate in blood plasma are important in clinical situations, because they are now known to be the spontaneous chemical products of nitric oxide (NO) in blood (*1*). NO is a mediator with numerous functions, including the regulation of vascular tone (*2*) and cell-to-cell communication (*3*). It is also thought to be one of the chief pathogenic factors causing shock and organ dysfunction seen in critically ill patients (*4*). Because NO is rapidly oxidized to nitrite and nitrate in blood, these are used as the markers of NO generation.

The most widely used method for determining nitrite and nitrate is based on the Griess reaction (*5*). However, this method requires pretreatment of the samples and the reduction of nitrate to nitrite prior to analysis. The pretreatment results in the contamination of sample, because of the prevalence of nitrite and nitrate in the environment.

Capillary electrophoresis (CE) enables the determination of nitrite and nitrate directly, with a minimum of pretreatment (*6–8*). The method we have developed here utilizes a detection wavelength of 214 nm, at which the high concentration of chloride ion in blood does not interfere with the quantification of nitrite and nitrate (*9,10*). Furthermore, the analysis can be completed in 20 min, so that CE may be useful for the determination of these anions in critical care medicine.

2. Materials
2.1. Reagents

1. All solutions should be prepared using 18-MΩ water generated by a Milli-Q laboratory water purification system (Millipore, Bedford, MA).

From: *Methods in Molecular Medicine, Vol 27: Clinical Applications of Capillary Electrophoresis*
Edited by: S. M. Palfrey © Humana Press Inc., Totowa, NJ

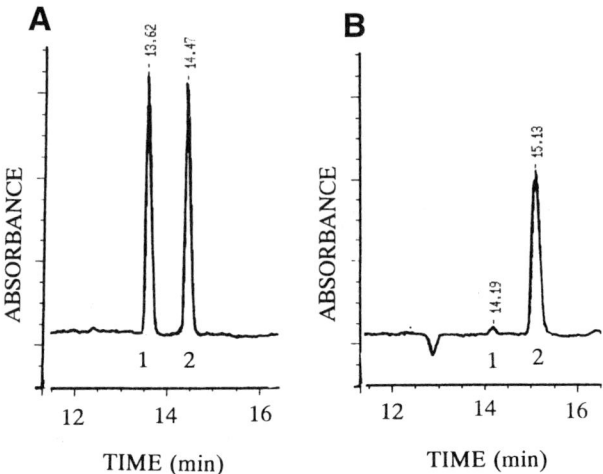

Fig. 1. Electropherogram of nitrite and nitrate. (**A**) Pattern of standard solution of nitrite and nitrate at individual concentration of 5 mg/L. (**B**) Representative pattern of normal human blood plasma. Peaks; 1, nitrite; 2, nitrate.

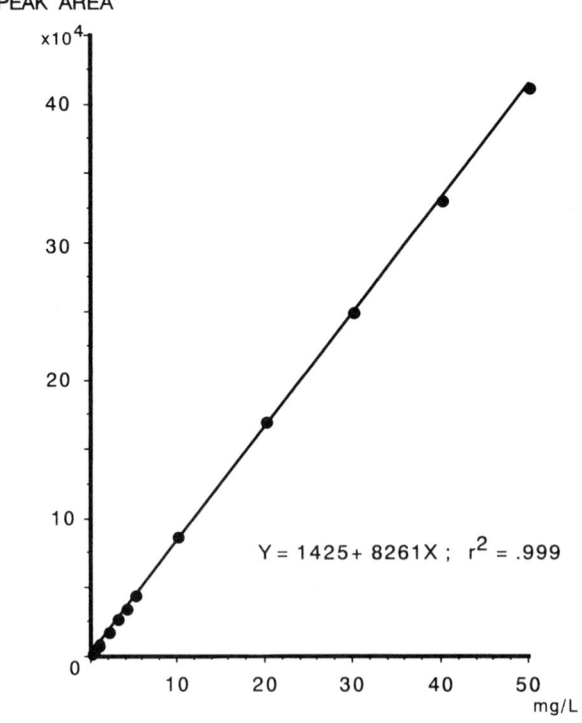

Fig. 2. Linearity of nitrite in the range from 0.1–50 mg/L.

2. Nitrite ion standard solution (1 mg/mL; Wako Pure Chemical Industries, Okasa, Japan) and nitrate ion standard solution (1mg/mL; Wako).
3. Working standards: Dilute 500 μL of each stock standard to 100 mL with 18-MΩ water.
4. Electrophoresis buffer: Dissolve 43.8 g sodium chloride and 50 mL NICE-Pak OFM (Waters, Milford, MA) in 18-MΩ water, and make up to 1000 mL (*see* **Note 1**). This solution is stable for 2 wk at –80°C.

2.2. Equipment

1. For the electrophoresis, the authors used a capillary ion analyzer (Waters ,Milford,MA) with UV filter (214 nm).
2. Model 805 data station (Waters,Milford, MA)).
3. Fused-silica capillary, 60 cm × 75 μm id.
4. 10 kDa mol wt cutoff centrifugal ultrafilter (UFC3, Millipore).

3. Methods
3.1. Capillary Preparation

1. Cut the capillary to a length a few centimeters longer than 60 cm.
2. Remove the coating for the detector window by burning a window approx 3–6 mm wide, 5–6 cm from one end, with a tool intended for burning a window in a capillary. Alternatively, a propane-type cigarette lighter can be used.
3. With a lint-free tissue, remove carbon particles from the treated area. Examine the window against a white background to ensure that all of the carbon has been removed.
4. Insert the capillary in the capillary holder soon after burning the window. The cathode is at the inlet end of the capillary, and the anode at the outlet end.
5. With a new capillary, rinse with 1 *M* sodium hydroxide for 20 min, followed by 18 –MΩ water for 20 min.

3.2. Preparation of Plasma Samples

1. Prepare human blood plasma by centrifuging heparinized blood at 1200*g* for 5 min. Two hundred μL of plasma is enough for the separation. Store the plasma at –80°C until use.
2. Deproteinize the plasma by centrifugal ultrafiltration (10 kDa mol wt cutoff), to yield 50 μL of sample for duplicated analyses (*see* **Note 2**).

3.3. Electrophoresis

1. Sample load time: 90 s hydrostatic loading.
2. Run time, 16 min.
3. Run voltage, 20 kV.
4. Temperature, 20°C.
5. Capillary purge: 2 min with 1 *M* sodium hydroxide, 2 min with 18–MΩ water, then 2 min with running buffer prior to each run (*see* **Note 3**).

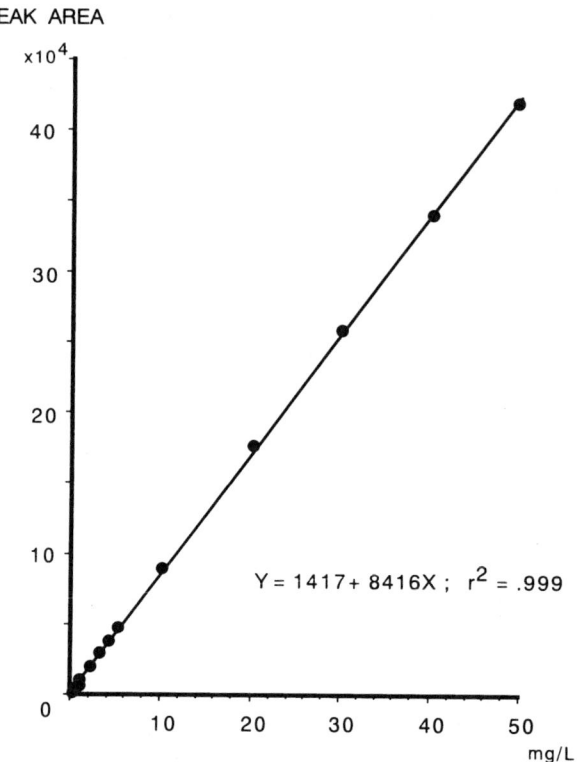

Fig. 3. Linearity of nitrate in the range from 0.1–50 mg/L.

6. Put 20 μL of the standard into vials 1 and 2. 20 μL of each sample can go into vials 3–16.
7. Put 1 M sodium hydroxide into vial 19, and 18–MΩ water into vial 20 for rinsing.
8. If the separation is to be done overnight, put 18–MΩ water into vial 17 to rinse the capillary after sample separation.
9. Separation of the two anions is usually complete within 14 min (**Fig. 1**). Check the migration time on the first sample, because the migration times are sometimes longer than this (*see* **Notes 4** and **5**).
10. In this method, the concentration and the peak area are linear for nitrite and nitrate in the range of 0.1 to 50 mg/L (**Figs. 2** and **3**).

4. Notes

1. Reagents for CE should be filtered through a 0.5 μm filter to remove small particles. This pretreatment may give more reproducible results.
2. Samples are easily contaminated, because nitrite and nitrate are prevalent in all laboratory ware, and especially in saliva *(6)*. It is good practice to wear a mask to avoid contaminating the sample with saliva.

3. The most common problem with this technique is an unstable baseline caused by residual proteins, bubbles, or stains inside the capillary. When the baseline becomes too unstable to read the peaks, rinse the capillary with 1 *M* sodium hydroxide for 20 min, and with 18–MΩwater for 20 min. If the baseline is not stabilized by rinsing the capillary, replace it with a new capillary.
4. Although the migration times of the two anions are usually 14 min (**Fig. 1**) this should be checked on the first sample.
5. If the migration time becomes longer than 16 min, check the ionic strength of the running buffer, and the analytical temperature.

References

1. Wennmalm, A., Benthin, G., and Petersson, A., S. (1992) Dependence of the metabolism of nitric oxide (NO) in healthy human whole blood on the oxygenation of its red cell haemoglobin. *Br. J. Pharmacol.* **106,** 507–508.
2. Palmer, R. M. J., Ferrige, A., and Moncada, S. (1987) Nitric oxide release accounts for the biological activity of endothlium-derived relaxing factor. *Nature* **327,** 524–526.
3. Moncada, S., Palmer, R. M. J., and Higgs, E. A. (1991) Nitric oxide: physiology, pathophysiology and pharmacology. *Pharmacol.Rev.* **43,** 109–142.
4. Ochoa, J. B., Udekwu, A. O., Billar, T. R., Curran, R. D., and Cerra, F. B., Simmons, R. L., and Peitzman, A. B. (1991) Nitrogen oxide levels in patients after trauma and during sepsis. *Ann. Surg.* **214,** 621–626.
5. Green, L. C. Wagner, D. A. Glogowski, J., Skipper, P. L., and Wishnok, J. S. Tannenbaum,S. R. (1982) Analysis of nitrate,nitrite, and [^{15}N] nitrate in biological fluids. *Anal. Biochem.* **126,** 131–138.
6. Romano, J., Jandik, P., Jones, W. R., and Jackson, P. E. (1991) Optimization of inorganic capillary electrophoresis for the analysis of anionic solutes in renal samples. *J.Chromatogr.* **546,** 411–421.
7. Jandik, P. and Jones, W. R. (1991) Optimization of detection sensitivity in the capillary electrophoresis of inorganic anions. *J. Chromatogr.* **546,** 432–443.
8. Wildman, B. J., Jackson, P. E., Jones, W. R., and Alden, P. G. (1991) Analysis of anion constitiuents of urine by inorganic capillary electrophoresis. *J. Chromatogr.* **546,** 459–466.
9. Leone, A. M., Francis, P. L., Rhodes, P., and Moncada, S. (1994) A rapid and simple method for the measurement of nitrite and nitrate in plasma by high performance capillary elctrophoresis. *Biochem. Biophys. Res. Commun.* **200,** 951–957.
10. Ueda, T., Maekawa, T., Sadamitsu, D., Oshita, S., Ogino, K., and Nakamura, K. (1995) The determination of nitrite and nitrate in human blood plasma by capillary zone electrophoresis. *Electrophoresis* **16,** 1002–1004.

Index

A

Adenylosuccinte lyase deficiency, 5
Alkaline phosphatase, 59
Amphetamine, 170
Ampholytes, 86, 96
Angiotensin convertase, 60
Anticonvulsants, 5, 153, 159
Apo proteins,
    A-I, 99, 100, 106
    A-II, 99, 100, 106
    J, 106
    B100, 99, 100, 105, 106
    C-I, 106
    E, 99
Apo-E gene, 123

B

β-Galactosidase, 59
BamHI, 66, 69, 71, 77
Bence Jones protein, 21, 22, 25, 26
Benzoylecgonine, 172
Bromide, 159
Buffers,
    effect of composition on EOF, 3, 4
Buffers, 3
    acetate, 40, 154
    bicarbonate, 40
    borax, 29, 167
    boric acid, 11, 13, 22, 48, 54, 61,
        104, 122, 159, 167
    formic acid, 61
    morpholinoethane sulfonic acid, 182
    quinine sulfate, 191, 196
    TBE, 68, 111, 129, 141
    TLE, 67, 111

C

C-reactive protein, see CRP
C3, 15
(CA)n repeats, 139, 142, 144
Capillaries,
    coated, 1, 22, 56, 68, 85, 113, 117,
        130, 141, 147, 183, 194, 197
    fused-silica, 2, 12, 22, 48, 53, 61, 104,
        122, 154, 159, 167, 191, 200, 205
    uncoated, 1, 104
Capillary gel electrophoresis, 66
Capillary isoelectric focusing, 4, 82
Capillary zone electrophoresis, 4, 47,
    61, 113, 115, 127
Cathepsin D, 61
CGE, see Capillary gel electrophoresis
Chronic hepatitis, 47
cIEF, see Capillary isoelectric focusing
Citrate, 5,200
CMV, see Cytoemegalovirus
Codeine, 168
Colorectal cancer, 144
Congenital adrenal hyperplasia, 184
Cortisol, 184
CRP, 15, 40, 41
Cryoglobulins, 5, 47, 48, 49
CSF, 5,29
Cystic fibrosis, 109
Cytomegalovirus, 65
CZE, see Capillary zone electrophoresis

D

11-deoxycortisol, 184
Detectors,
    diode array, 2, 105, 168

laser-induced fluorescence, 2, 66, 72, 118, 121, 122, 125, 129, 141
UV absorption, 2, 22, 113, 128, 130, 183, 205
DNA,
  amplification, 66
  carrier, 70
  CMV, 71, 77
  double-stranded, 121, 122, 127
  extraction, 67, 111, 112
  genomic, 123, 141
  ladder, 113, 141, 148
  plasmid, 66
  single-stranded, 127, 132, 133, 135
  viral, 66
Down's syndrome, 5, 110, 113
Drugs,
  of abuse, 5
Duchenne muscular dystrophy, 109

E

Ecstasy, *see* MDMA
EDDP, 170
Endosmotic flow, 3, 33, 105, 117, 122, 166, 185
EOF marker, 107
EOF, *see* Endosmotic flow
Ethidium bromide, 117, 121, 122

F

Fragile-X syndrome, 139
Free solution capillary electrophoresis, 29, 39

G

Glomerular proteinuria, 21, 24
Glucose-6-phosphatase, 60
Glucose-6-phosphate dehydrogenase, 59
Glutathione peroxidase, 60

H

HbA$_1$c, 5
HDL, 99, 100, 104, 106

Hemoglobin Bart's, 97
Hemoglobin-C-Harlem, 94
Hemoglobin H, 97
Hemoglobin-O-Arab, 94
Hemoglobin variants, 5,81
Hepatitis C, 110
Hereditary nonpolyposis colorectal cancer, 139
Hereditary persistence of fetal hemoglobin, 92
*Hin*dIII, 78
HIV-1 virus, 110
Huntington's disease, 139
Hydrocodeine, 168
Hydroxyethylcellulose, 117
17-hydroxyprogesterone, 184
Hydroxypropylmethylcellulose, 68, 150

I

Immunoalobulin A, 41
Immunoglobulin G, 41
Immunoglobulin M, 41
Immunosubtraction, 39
Indirect UV detection, 190, 199
Insulin, 159
Iohexol, 160
Isoelectric focusing, *see* Capillary isoelectric focusing
Isotachophoresis, 5, 118, 159

K

Kappa light chains, 41

L

Lambda light chains, 41
Lamotrigine, 153,154
Laser-induced fluorescence, 66, 72, 118, 121, 122, 125, 144
LDL, 99, 100, 104, 106
LIF, *see* Laser-induced fluorescence
Lipoproteins, 5

M

Macroglobulinaemia, 39, 47
MDA, 170
MDE, 170
MDMA, 170
MECC, *see* Micellar electrokinetic
  capillary chromatography
Methadone, 170
Methamphetamine, 170
Methyl cellulose, 22, 85, 128, 140
Micellar electrokinetic capillary
  chromatography, 5, 185
Microsatellite instability, 139, 144
Monoclonal proteins, 12, 14, 39, 41, 49
Morphine, 168
Myeloma, 39, 47
Myoglobin, 53

N

Nitrate, 5, 159, 203
Nitric oxide, 203
Nitrite, 5, 205
Nongel-sieving capillary
  electrophoresis, 128, 133, 139,
  140, 146
Non-Hodgkin's lymphoma, 39

O

Oxalate, 5, 200

P

p53, 5, 127, 135
Paraproteins, *see* Monoclonal proteins
PCR products,
  desalting of, 73, 112, 113, 117
PCR, *see* Polymerase chain reaction
Phenobarbital, 159
Phenytoin, 160
Point mutation, 130
Polio virus, 110
Polyacrylamide-filled capillaries,
  112, 117, 121

Polymerase chain reaction, 67, 110,
  111, 112, 122, 149
Procainamide, 159
Putrescine, 189

Q

Quinidine, 159

R

Raynaud's phenomena, 47
Rhesus D/d genotyping, 110
Rheumatoid arthritis, 47

S

Sample desalting, 73, 112
Sample loading, 2
  electrokinetic, 2, 72, 168, 173,
    192, 194
  hydrodynamic, 2, 3, 30, 48, 54, 61,
    105, 130, 149, 154, 160, 168,
    183, 200, 205
Sepharose, 40
Serum proteins, 5, 11, 39
Sickle cell disease, 91
Sieving liquid polymers, 117, 121
Single-strand conformation
  polymorphism analysis, 127,
  132, 135
Sjögren's syndrome, 47
Sodium chromate, 199
Solid-phase extraction, 100, 104, 166,
  167, 181, 185
SPE, *see* Solid-phase extraction
Spermidine, 189
Spermine, 189
SSCP*, see* Single-strand conformation
  polymorphism analysis
Stacking, 157, 201
Stacking,
  acetonitrile, 158
Sudan black B, 103
Surfactants,

n-alkyldimethylamino-propane
        sulfonate, 54
SB3-12, 185
SDS, 100, 104, 105, 186
Tetra-decyl-trimethylammonium
        bromide, 199
SYBR green I, 68
Systemic lupus erythematosus, 47

T

Tau protein, 32
Thalassemias, 81, 91, 92
Transferrin, 40
Triple-X syndrome, 110

Trisomy, 18, 110
Trisomy, 21,110
Tubular proteinuria, 21, 24
Tyramine, 154

U

Urine free light chains, 46
Urine protein, 5, 21, 39
Urine steroids, 5

V

Viral load, 65, 75
VLDL, 99, 100, 104